A2 UNIT 1

STUDENT GUIDE

T0187374

CCEA

Biology

Physiology, coordination and control, and ecosystems

John Campton

HODDER
EDUCATION
AN HACHETTE UK COMPANY

Hodder Education, an Hachette UK company, Carmelite House, 50 Victoria Embankment, London EC4Y 0DZ

Orders

Please contact Hachette UK Distribution, Hely Hutchinson Centre, Milton Road, Didcot, Oxfordshire, OX11 7HH

tel: 01235 827827

e-mail: education@hachette.co.uk

Lines are open 9.00 a.m.–5.00 p.m., Monday to Friday. You can also order through the Hodder Education website: www.hoddereducation.co.uk

© John Campton 2017

ISBN 978-1-4718-6303-5

First printed 2017

Impression number 7

Year 2022

All rights reserved; no part of this publication may be reproduced, stored in a retrieval system, or transmitted, in any other form or by any means, electronic, mechanical, photocopying, recording or otherwise without either the prior written permission of Hodder Education or a licence permitting restricted copying in the United Kingdom issued by the Copyright Licensing Agency Ltd, www.cla.co.uk.

This guide has been written specifically to support students preparing for the CCEA A-level Biology examinations. The content has been neither approved nor endorsed by CCEA and remains the sole responsibility of the author.

Cover photo: andamanse/Fotolia; p.80, Don Fawcett/SPL; p.84 Biology Media/SPL.

Typeset by Integra Software Services Pvt. Ltd, Pondicherry, India

Printed and bound by CPI Group (UK) Ltd, Croydon, CR0 4YY

Contents

Content Guidance

Questions & Answers

■ Getting the most from this book

Exam tips

Advice on key points in the text to help you learn and recall content, avoid pitfalls, and polish your exam technique in order to boost your grade.

Knowledge check

Rapid-fire questions throughout the Content Guidance section to check your understanding.

Knowledge check answers

1 Turn to the back of the book for the Knowledge check answers.

Summaries

■ Each core topic is rounded off by a bullet-list summary for quick-check reference of what you need to know.

Exam-style questions

Commentary on the questions

Tips on what you need to do to gain full marks, indicated by the icon **e**

Sample student answers

Practise the questions, then look at the student answers that follow.

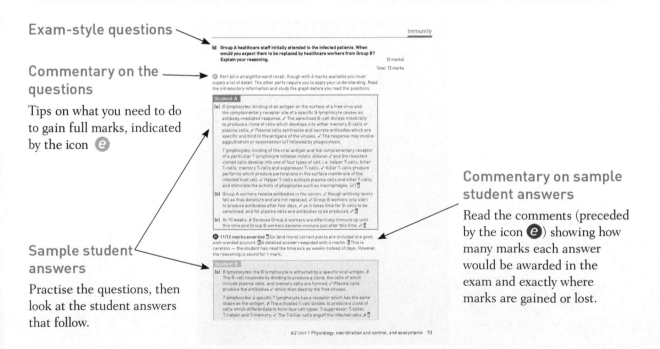

Commentary on sample student answers

Read the comments (preceded by the icon **e**) showing how many marks each answer would be awarded in the exam and exactly where marks are gained or lost.

■About this book

The aim of this book is to help you prepare for the A2 Unit 1 examination for CCEA Biology.

The **Content Guidance** section contains everything that you should learn to cover the specification content of A2 Unit 1. It should be used as a study aid as you meet each topic, when you prepare for end-of-topic tests, and during your final revision. For each there are *exam tips* and *knowledge checks* in the margins. Answers to the knowledge checks are provided towards the end of the book. At the end of each topic there is a list of the *practical work* with which you are expected to be familiar. This is followed by a comprehensive, yet succinct, *summary* of the points covered in each topic.

The **Questions & Answers** section contains questions on each topic. There are answers written by two students, together with comments on their performances and how the answers might have been improved. You will encounter a range of question styles in the A2 Unit 1 exam, and the students' answers and comments should help with your examination technique.

Developing your understanding

It is important that throughout your A-level course you develop effective study techniques.

- You must *not* simply read through the content of this book.
- Your understanding will be better if you are *active* in your learning. For example, you can take the information given in this book and present it in different ways:
  - a series of bullet points summarising the key points
  - an annotated diagram to show structure and function, e.g. a diagram of a kidney nephron with labelled features and noting what happens in each region with respect to filtration and reabsorption
  - an annotated graph, e.g showing the phases of population growth, with notes explaining what is happening in each phase
  - a spider diagram, e.g. one on the nerve impulse would include reference to neurone structure; the resting potential, the action potential and its propagation as an impulse, and the factors that increase transmission speed
- Compile a glossary of terms for each topic. Key terms in this guide are shown in **bold** (with some defined in the margin) and for each you should be able to provide a definition. This will develop your understanding of the language used in biology and help you where *quality of written communication* is being assessed.
- Write essays on different topics. For example, an essay on the immune response will test your understanding of the entire topic and give you practice for the Section B question.
- *Think* about the information in this book so that you can *apply* your understanding in unfamiliar situations. Ultimately you will need to be able to deal with questions that set a topic in a new context.

- Ensure that you are familiar with the expected practical skills, as questions on these may be included in this unit.
- Use past questions and other exercises to develop all the skills that examiners must test.
- Use the topic summaries to check that you have covered all the material you need to know and as a brief survey of each topic.

Synoptic links

In order to develop your understanding of the subject as a whole, you need to work at making the connections between the topics you have studied so far. It is essential that you revisit the core concepts that you learned at AS since these often underlie the topics at A2. For example, many areas of biology rely on an understanding of cell biology (ultrastructure and function) since cells are the 'units of life'. Other synoptic links are highlighted in the Content Guidance.

Content Guidance

■ Homeostasis and the kidney

The kidneys and excretion

The kidneys regulate the internal environment by constantly adjusting the composition of the blood. They are the organs of excretion and osmoregulation.

Excretion is the removal from the body of the toxic waste products of metabolic processes. In mammals, carbon dioxide, produced during respiration, is excreted from the lungs. The kidneys excrete nitrogen-containing compounds, mostly **urea**, produced during the breakdown of excess amino acids and nucleic acids in the liver. The kidneys also excrete a little creatinine, a waste product produced from the degradation of creatine phosphate (a molecule of major importance in ATP generation) in muscles.

Homeostasis is the maintenance of steady states within the body. The kidneys have a homeostatic function in regulating the water content of the blood. The kidneys control the water potential of body fluids (**osmoregulation**) under the influence of antidiuretic hormone.

The structure of the urinary system

The components of the urinary system of a mammal are shown in Figure 1.

Excretion The removal of waste products of metabolism from the body.

Exam tip

It is important to distinguish the terms *excretion* and *egestion*. Egestion is the removal from the body of waste material, such as undigested food, which has not been part of the body's metabolism, i.e. faeces are not excreted, they are egested.

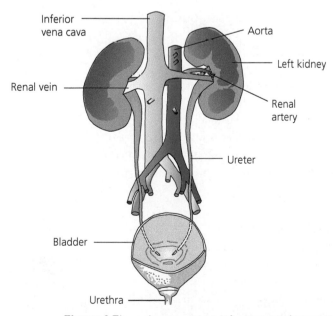

Figure 1 The urinary system of a mammal

An indication of the importance of the kidneys in regulating the composition of the blood is the fact that they receive approximately 25% of the cardiac output (via the aorta and renal arteries).

Inside a kidney there are two layers — an outer **cortex** and an inner **medulla** — surrounding a central cavity, the **pelvis**. The medulla is sub-divided into a number of **pyramids** whose apices protrude into the pelvis. A kidney contains more than 1 million microscopic tubules called **nephrons**, each of which has a rich blood supply. The positioning of nephrons relative to other regions of the kidney is shown in Figure 2(a). The overall structure of a nephron and its blood supply is shown in Figure 2(b).

> **Exam tip**
>
> You are expected to be able to recognise the cortex, medulla, pyramid, pelvis, ureters, bladder and urethra in photographs or diagrams.

> **Exam tip**
>
> You may be expected to interpret the histology of the kidney. Make sure that you use the information provided about the region of the kidney from which the section is taken. For example, if it is from the cortex you might expect to see the glomerulus and Bowman's capsule or, alternatively, sections of the convoluted tubules.

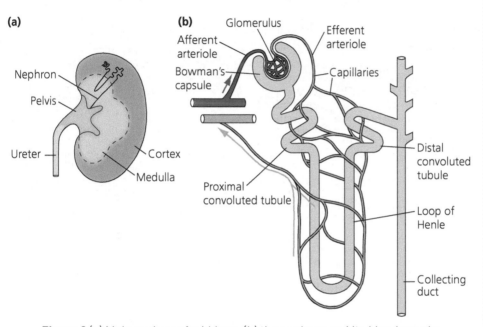

Figure 2 (a) Main regions of a kidney; (b) the nephron and its blood supply

The nephron is the functional unit of the kidney. Each nephron consists of a cup-shaped **Bowman's capsule**, plus a tube that has three distinct regions: the **proximal (first) convoluted tubule**, the **loop of Henle** and the **distal (second) convoluted tubule**. Arterial blood enters each capsule through an **afferent arteriole** which branches to form a capillary network called the **glomerulus**; blood leaves the glomerulus through an **efferent arteriole** which then branches to form a further capillary network (the vasa recta system) around the main body of the nephron. A number of nephrons join to form a **collecting duct** which transfers the fluid towards the pelvis.

The production of urine

Urine is produced in two stages involving quite different processes:

- **Ultrafiltration** is the stage in which plasma in the glomerulus is filtered into Bowman's capsule. Only substances below a certain size are filtered and so the filtrate contains useful molecules as well as toxic ones.
- **Reabsorption** of useful substances back into the blood occurs as the filtrate passes along the nephron and collecting duct. Only at the point where the collecting duct joins with the pelvis can the fluid be called 'urine'.

> **Knowledge check 1**
>
> Suggest why nephrons have tubules with convolutions.

Ultrafiltration

The driving force

The blood entering the glomerulus is under high pressure. This high **hydrostatic pressure** occurs because:

- the renal arteries are wide, short and relatively close to the heart
- the efferent arteriole is smaller than the afferent arteriole (carrying blood to the glomerulus), which creates a bottleneck

It is this high hydrostatic pressure which effectively causes fluid to filter from the glomerular plasma as filtrate in the capsule. Ultrafiltration is so efficient that 15–20% of the water and solutes are removed from the plasma that flows through the glomeruli.

The filter

There are only three layers separating plasma from filtrate. These are the capillary endothelium, the basement membrane on which the capillary cells lie and the inner layer of the Bowman's capsule. Two of these layers are especially porous:

- the endothelium of the capillaries in the glomerulus, which consists of a single layer of squamous (thin) cells with pores between them
- the inner wall of the Bowman's capsule consists of podocytes, with foot-like processes which surround the capillaries but which have spacious gaps between them called filtration slits

The *effective filter* is the **basement membrane** of the glomerular capillaries. This extracellular membrane lies on the outer side of the capillary endothelium. These layers are shown in Figure 3 (not drawn to scale).

Knowledge check 2

What creates the high hydrostatic pressure in the glomerulus?

Knowledge check 3

Explain how the structure of (a) the endothelial cell of the glomerular capillaries, (b) the basement membrane and (c) the podocytes influences their roles in ultrafiltration.

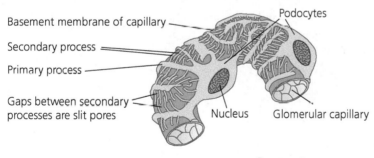

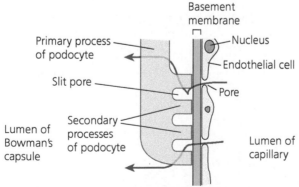

Figure 3 The structure of the filter

The composition of the filtrate

Normally only molecules with a relative molecular mass of less than 68 000 can pass through the basement membrane. All constituents of the blood plasma other than plasma proteins — except for the smallest of these — are able to pass through. The filtrate consists mainly of inorganic ions, glucose, amino acids, urea and other toxic molecules, all dissolved in water. Clearly, water and useful substances must be reabsorbed.

Synoptic links

Osmosis and water potential

For a fuller understanding of ultrafiltration and reabsorption you need to know that water moves from a region of high water potential to a region of lower water potential, and water potential has two components: the solute and pressure potentials (from AS Unit 1).

Calculation of the net filtration pressure between glomerulus and capsule

Overall, the water potential of the glomerular plasma exceeds the water potential of the filtrate in the capsule. This is due to the large pressure potential (high hydrostatic pressure) within the glomerulus. This is, in part, opposed by a more negative solute potential within the plasma (in which the retained proteins act as solutes, though they are more accurately colloids in solution) than in the filtrate, while there is some resistance to further filtration due to back pressure of the filtrate in the nephron.

The net filtration force is the difference in water potential either side of the filter, i.e. water potential of glomerular plasma minus water potential of filtrate in the capsule. Net filtration force is calculated as follows:

$$\Psi_{plasma} = \Psi_s \text{ (due to proteins)} + \Psi_p \text{ (hydrostatic pressure)}$$

$$= (-3.3\,\text{kPa}) + 6\,\text{kPa}$$

$$= 2.7\,\text{kPa}$$

$$\Psi_{filtrate} = \Psi_s \text{ (no proteins present)} + \Psi_p \text{ (hydrostatic pressure)}$$

$$= 0 + 1.3\,\text{kPa}$$

$$= 1.3\,\text{kPa}$$

Therefore:

$$\text{net filtration force} = \Psi_{plasma} - \Psi_{filtrate} = 1.4\,\text{kPa}$$

Knowledge check 4

What is the concentration of glucose in the glomerular filtrate relative to its concentration in the blood plasma?

Reabsorption

Reabsorption in the proximal convoluted tubule

As the filtrate flows through the proximal convoluted tubule, 80% of the water is reabsorbed by osmosis into adjacent blood capillaries, while 67% of ions are reabsorbed — partly by diffusion following the reabsorption of water and partly by active transport. All the glucose and amino acids pass back into the blood by active transport. Small proteins, which may have been filtered, are reabsorbed by pinocytosis. By the end of the proximal convoluted tubule, the filtrate is isotonic with the plasma (i.e. it has the same water potential).

The cuboidal epithelial cells, which line the tubule walls, have numerous **microvilli** on the luminal surface and **infoldings** of the basal cell-surface membrane next to the blood capillaries. These adaptations greatly increase the surface area available for reabsorptive processes. Furthermore, the cells have many **mitochondria** located near the infoldings and these supply the extra ATP needed for active transport (see Figure 4).

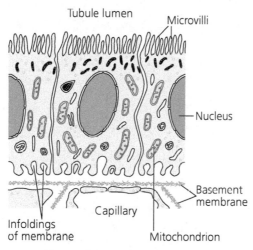

Figure 4 A proximal tubule cell

Synoptic links

Movement across the cell-surface membrane

In AS Unit 1 you learned about diffusion, facilitated diffusion, active transport and pinocytosis.

The role of the loop of Henle

Sodium (Na^+) and chloride (Cl^-) ions pass out of the ascending limb (by a combination of diffusion and active transport) into the surrounding tissue of the medulla (Figure 5), lowering its water potential. This creates an osmotic gradient which draws water out of the permeable descending limb, to be carried away by blood in the surrounding capillaries (the ascending limb is impermeable to water).

Knowledge check 5

Explain why reabsorption from the nephron must be selective, i.e. some molecules are reabsorbed more readily than others.

Exam tip

The cells of the proximal tubule possess a range of adaptations to increase the efficiency of reabsorption: microvilli increasing surface area; infoldings of membranes; abundance of mitochondria; close association of many capillaries; thin squamous endothelium of capillaries; flow of filtrate and blood maintaining concentration gradients.

This combination of water and ion movement causes fluid (in and around the loop) to be saltier as it goes down the descending limb and become less salty as it goes up the ascending limb. This means that the tissues deeper in the medulla have a higher concentration of ions and lower (more negative) water potential. As the collecting ducts pass through the medulla (and assuming they are permeable to water), water is reabsorbed by osmosis, resulting in a more concentrated urine.

Exam tip

The wall of the upper thick region of the ascending limb is also composed of cuboidal cells rich in mitochondria. The mitochondria provide the ATP necessary to pump Cl^- ions into the medulla.

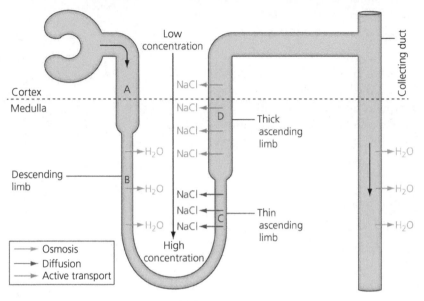

Figure 5 The loop of Henle

Exam tip

The longer the loop of Henle, the greater is the decrease in water potential within the medulla and so the more water can be reabsorbed. Desert animals, such as the kangaroo rat and the camel, have very long loops of Henle as an important adaptation enabling them to conserve water.

Reabsorption in the distal convoluted tubule

In the distal convoluted tubule the ionic composition and the pH of the blood are adjusted. It is also here that toxic substances, such as creatinine, are secreted into the filtrate. Depending on the permeability of the tubule walls, some water may be reabsorbed.

Osmoregulation

The collecting ducts are where the water content of the blood (and therefore of the whole body) is regulated. The permeability to water of the walls of the second convoluted tubule and collecting duct is increased by **antidiuretic hormone** (ADH). ADH is *produced* by the **hypothalamus** but *secreted* into the **posterior lobe of the pituitary body**, where it is stored. A rise in blood concentration (i.e. when the water potential of the blood becomes more negative) is detected by **osmoreceptors** in the hypothalamus. These receptors send impulses to the posterior lobe of the pituitary gland. As a result, this lobe of the pituitary gland releases *more ADH* into the blood which *increases the permeability* to water of the second convoluted tubule and collecting duct. In fact, water moves through channel proteins (aquaporins) which open, under the influence of ADH, to let water through. More water passes to the medulla and a more concentrated (hypertonic) urine is produced.

Knowledge check 6

Name (a) the structure responsible for the detection of changes in the water potential of blood, and (b) the gland responsible for the release of ADH.

A fall in blood concentration (i.e. the water potential of the blood becomes less negative) *inhibits the release of ADH*. As a result, the walls of the second convoluted tubule and collecting duct become *impermeable to water*, less water is reabsorbed and a less concentrated (hypotonic) urine is produced.

Osmoregulation involves **negative feedback**: it is *feedback*, since a change in the water potential of the blood (detected by the osmoreceptors and determining the release of ADH) will ultimately lead to another change in the water potential of the blood; it is *negative*, since an increase in water potential (e.g. blood diluted by drinking) will later result in a decrease in water potential (see Figure 6).

Knowledge check 7

Is a dehydrated person likely to produce more or less ADH?

Knowledge check 8

Suggest why beavers have short loops of Henle.

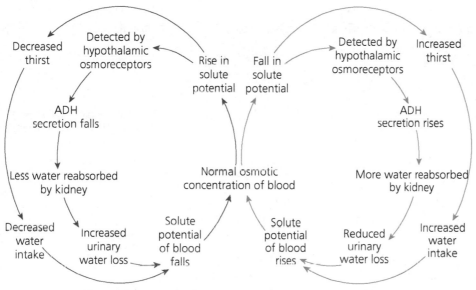

Figure 6 Osmoregulation as a negative feedback process

Summary

- Excess amino acids in the diet are deaminated (in the liver), producing urea, which enters the blood circulatory system to be transported to the kidneys.
- The kidney is composed of about 1 million small tubules, called nephrons, organised within distinct layers — the outer cortex and the inner medulla.
- Each nephron is supplied with blood vessels, which form a knot of capillaries, the glomerulus, enclosed by the cup-shaped Bowman's capsule, and which surround all other parts of the nephron.

- Each nephron produces urine as follows:
  - Ultrafiltration takes place from the glomerulus into the Bowman's capsule. Small molecules are filtered, by the basement membrane of the glomerulus endothelium, under pressure. The glomerular filtrate is hypotonic to the blood as it lacks plasma proteins.
  - In the proximal convoluted tubule, glucose, amino acids and some ions are actively transported back into the blood during selective reabsorption. Water follows by osmosis and the filtrate becomes isotonic to the blood.

→

— The movement of solutes from the ascending limb of the loop of Henle (and reabsorption of water from the descending limb) creates a region of high solute concentration (low water potential) gradient deep in the medulla. This enables water to be withdrawn from the collecting ducts if they are permeable to water.

— Before the filtrate passes down the collecting duct, it travels through the distal convoluted tubule where further ions are actively reabsorbed.

■ Antidiuretic hormone (ADH) makes the wall of the distal convoluted tubule and collecting ducts permeable to water by causing water channel proteins, called aquaporins, to open.

■ ADH is produced in the hypothalamus but stored in the posterior lobe of the pituitary gland from where it is secreted into the blood when hypothalamic osmoreceptors detect that the blood water potential has fallen below the norm. ADH enables water to be reabsorbed through the tubule walls so that hypertonic urine is produced.

■ If the water potential of the blood rises, then ADH release is reduced and the impermeability of the tubule walls prevents the reabsorption of water. More water passes out in the dilute urine and the water potential of the blood falls.

■Immunity

Pathogens and the spread of disease

The conditions in the body provide an ideal environment for the growth of a vast array of microorganisms. These include viruses, bacteria, fungi and protoctists, which are present in the air we breathe, in the water and food we consume and on the objects we touch. However, relatively few of these microorganisms are disease-causing, or **pathogenic**.

For a microorganism to be pathogenic it must (1) enter the host, (2) colonise tissues of the host, (3) evade the host's defences and (4) cause damage to the host tissues. If a pathogen gets into the host and colonises its tissues, an **infection** results. **Disease** occurs when an infection leads to recognisable symptoms in the host. The disease may cause damage to host tissues directly (e.g. viruses cause cells to break down) or through the production of toxins (e.g. as many bacteria do). Examples of some infectious or communicable diseases are given in Table 1.

Knowledge check 9

Describe the differences between a parasite and a pathogen.

Exam tip

The infections listed in Table 1 are provided only so that you may be aware of the causes and effects of some examples. You do not have to learn them.

Table 1 Examples of infectious diseases caused by a variety of pathogenic microorganisms

Disease	Microorganism	Tissue infected	Symptoms	Transmission
Measles	An RNA virus of the genus *Morbillivirus*	Gas exchange system	Fever and a characteristic rash	Airborne transmission
Poliomyelitis ('polio')	An RNA virus of the genus *Enterovirus*	Nervous tissue	In the most severe cases paralysis occurs	Direct contact with an infected person
Acquired immune deficiency syndrome (AIDS)	Human immunodeficiency virus — a retrovirus of the genus *Lentivirus*	Helper T-cells of the immune system (specifically CD4+ T-cells) — see later	Occurrence of diseases normally controlled by the immune system that HIV damages	Transmitted through body fluids, e.g. semen, blood

→

Disease	Microorganism	Tissue infected	Symptoms	Transmission
Ebola virus disease (Ebola haemorrhagic fever)	Viruses (4) of the genus *Ebolavirus*	Macrophages, connective tissue and endothelial cells (e.g. of blood vessels)	Fever, muscle weakness followed by internal and external bleeding	Contact with body fluids or organs of an infected person; some species of fruit bat are considered to be reservoirs of the virus
Tetanus ('lockjaw')	The anaerobic bacterium *Clostridium tetani*	The bacteria produce a toxin that passes to all parts of the body	Prolonged contraction of skeletal muscle fibres	Infection may occur when dirt enters a deep puncture wound
Tuberculosis ('TB')	The bacterium *Mycobacterium tuberculosis*	The lungs	Coughing, shortness of breath, fever and extreme fatigue	Spread in microscopic droplets when a sufferer coughs or sneezes
Athlete's foot	A fungus of the genus *Trichophyton*	The upper layer of the skin (when moist and warm)	Itching of the skin, which may crack and peel	Contact with dead skin cells carrying the fungi which thrive in moist environments (e.g. showers)
Malaria	A protoctist of the genus *Plasmodium*	Red blood cells and liver tissue (where *Plasmodium* can remain dormant)	Anaemia, fever, chills, nausea and, in severe cases, coma and death	Transmitted by the *Anopheles* mosquito which acts as a vector (carrier)

Table 1 shows a number of ways in which infections may be spread (transmitted). The disease may be spread through a population at a local or national level (**epidemic**) or across whole continents and even worldwide (**pandemic**).

The body has three lines of defence to resist pathogens:

- The first is to prevent the entry of pathogens.
- If this fails, the second is for phagocytes to gather at the site of infection where they ingest pathogens.
- If this fails, the third is for the body to target that particular pathogen, in a specific immune response.

Barriers to pathogen entry

The barriers to the entry of pathogens are:

- the **skin**, the outer layer of which consists of layers of dead cells filled with a tough protein (keratin) and covered in an oily secretion (sebum)
- **tears** and **saliva**, which contain the enzyme **lysozyme**, capable of digesting the cell walls of many bacteria
- **mucus**, secreted by goblet cells lining the respiratory tract, which traps microorganisms and contains lysozyme, preventing them from penetrating the underlying membranes, while cilia normally sweep this mucus up and out
- **acid** (specifically hydrochloric acid), secreted in the stomach, which kills most of the bacteria entering in the water and food consumed

Nevertheless, pathogens can still enter the body, most readily through broken skin (e.g. as the result of a cut, though blood clotting will limit entry via this route) or via the large absorptive surfaces in the lungs and the intestines.

Epidemic An outbreak of an infectious disease that spreads rapidly among individuals of a population at the same time.

Pandemic An outbreak of a disease that occurs over a wide geographic area and affects an exceptionally high proportion of the population.

Knowledge check 10

Suggest how bacteria living in your intestines can help to prevent infection of the gut by pathogenic bacteria.

The activity of phagocytes

Synoptic links

Phagocytosis and lysosomes

In AS Unit 1 you learned about phagocytosis whereby small organisms, such as bacteria, are engulfed; and the role of lysosomes in intracellular digestion.

If a pathogen does get into the body, changes occur in the infected area to cause the capillaries to become leaky. As plasma escapes, the area becomes swollen (in what is known as an inflammatory response), but, more importantly, phagocytic white blood cells readily squeeze through the leaky capillary walls and accumulate in the area. These phagocytes include:

- **polymorphs**, which are the most common phagocytes and so the first to arrive
- **macrophages**, which develop from the monocytes in the blood and are larger and longer-lived

Both types of phagocyte engulf the bacteria and the debris from damaged cells. Ingested material is enclosed within a vacuole; then, lysosomes fuse with the vacuole, releasing hydrolytic enzymes which destroy the bacteria. The process is shown in Figure 7.

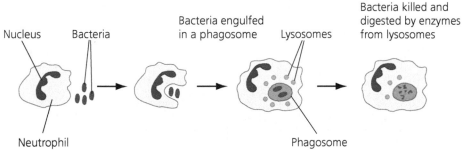

Figure 7 The activity of a phagocyte in engulfing and destroying bacteria

The specific immune response

Lymphocytes: the B-cells and T-cells

The **immune response** is a specific response to the detection of pathogens in the body. It involves **lymphocytes**, another type of white blood cell. The pathogens carry molecules on their outer surfaces which are recognised by the body as being **foreign** or '**non-self**'. These molecules may consist of protein, carbohydrate or glycoprotein. They initiate the immune response and are called **antigens**. Different pathogens have different antigens. Lymphocytes all look the same; however, they differ biochemically. For each antigen that occurs on the surface of a pathogen there is a lymphocyte that carries a special **protein receptor** on its cell-surface membrane which is complementary in shape. It is the complementary nature of the receptor and antigen (similar to the lock-and-key model of enzyme action) that ensures a specific response.

Exam tip

You should be able to compare the two types of phagocyte. Polymorphs, more abundant than monocytes though shorter-lived, remain in the blood until such time as there is a bacterial infection, when they move rapidly out of the blood capillaries to accumulate at the site of infection. Monocytes are blood cells that, prior to infection, can move out of the blood capillaries into the surrounding tissues (e.g. the alveoli of the lungs), where they develop into macrophages. Both are phagocytic.

Knowledge check 11

Explain why
lymphocytes do not
attack our own cells.

Synoptic links

Membrane structure

In AS Unit 1 you learned that the outer surface of the cell membrane contains proteins, carbohydrate and glycoproteins — it is these that act as antigens.

During early development, many millions of different lymphocytes are produced from the stem cells in the bone marrow. Each carries a specific membrane receptor that allows it to respond to a different non-self antigen if this should be encountered in the future. This gives the immune system the ability to respond to any type of pathogen that enters the body. There are no lymphocytes set up to respond to any molecules on the cell-surface membranes of that individual's own cells ('self' antigens).

In fact, there are two sets of lymphocytes: **B lymphocytes** and **T lymphocytes** (also called **B-cells** and **T-cells** respectively; see Table 2). Both types are produced in the bone marrow, but they differ in terms of where they mature and the nature of the immune response produced.

Both B and T lymphocytes migrate to lymphoid tissue throughout the body to await possible activation.

Stem cells Unspecialised cells that can give rise to a variety of specialized cells.

Exam tip

You must be able to distinguish B lymphocytes and T lymphocytes. It is not sufficient to know that B-cells are produced in **bone** marrow and that T-cells mature in the **thymus** gland. What is more important is that **B**-cells result in the production of anti**b**odies (so T-cells are associated with cell-mediated immunity).

Table 2 A comparison of B and T lymphocytes

Type of lymphocyte	Maturation	Type of immune response	Nature of immune response
B lymphocytes (B-cells)	Continue maturing in the **B**one marrow	Antibody-mediated (humeral) immunity	Secrete antibodies which counter the antigen-carrying pathogens
T lymphocytes (T-cells)	Mature in the **T**hymus	Cell-mediated immunity	Attack infected cells with the antigen presented on the surface

Activation of lymphocytes by 'non-self' antigens

Activation of a lymphocyte involves it coming into contact with a 'non-self' antigen that its receptor recognises. Remember, there may be only one or a few of the 'correct' B and T lymphocytes available. The appropriate B-cell will recognise the antigen on the pathogen itself, while the appropriate T-cell will recognise infected cells since they display the pathogen's antigens on their surface, i.e they act as **antigen-presenting cells**. Of course, it takes some time for the lymphocytes to come into contact with the antigen. Once they have made contact with the antigen, the lymphocytes become **sensitised**. In B-cells, this involves the activation of a gene for production of antibodies. The sensitised lymphocytes then divide by **mitosis** a number of times — they are **cloned** — and differentiate into a variety of cells. Again, this takes time. Overall, there is a delay of about four days, between contact with the antigen and the cloning of the required lymphocytes, during which time the person suffers from the disease caused by the pathogen.

Antibody-mediated immunity

Most of the cloned B-cells develop into **plasma cells** which synthesise and secrete large amounts of the **antibody**. Their activity is intense — they may produce several thousand antibodies per second — and so they are short-lived.

After several weeks their numbers decrease but the antibodies remain in the blood for some time. Eventually, however, the concentration of antibodies decreases too.

Antibodies are globular proteins called **immunoglobulins**. An antibody has a binding site with a shape complementary to a particular antigen (see Figure 8) and forms an antigen–antibody complex.

Synoptic links

Protein structure

In AS Unit 1 you learned that proteins have a specific structure — it is this specificity that allows the binding site to be complementary to a particular antigen.

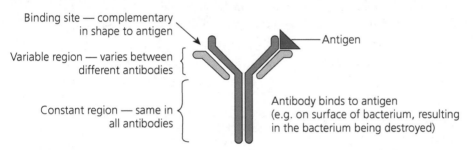

Figure 8 The structure of an antibody (binding to an antigen)

> **Exam tip**
>
> Don't confuse the terms *antigen* and *antibody*. Antigens are foreign (non-self), while antibodies are produced in the **body** by **B**-cells.

Antibodies function in a variety of ways. Some:
- may neutralise toxins produced by bacteria — these antibodies are called **antitoxins**
- clump or agglutinate bacteria before the latter are engulfed by phagocytic cells — these antibodies are **agglutinins**
- attach to viruses, preventing them from entering and infecting host cells
- destroy bacterial cell walls, causing lysis (the cells burst)
- attach to bacteria, enabling phagocytic cells to identify them

Some of the cloned B-cells remain in the blood but they are not plasma cells and do not secrete antibodies. They are **memory cells** and they live for a very long time. If the same antigen is encountered again, the memory cells rapidly clone to produce plasma cells which secrete antibodies. This rapid response means that no 'infection' is suffered on the subsequent occasion — the person has become immune to that particular disease.

The way in which antibody-mediated immunity defends against bacteria is illustrated in Figure 9.

1 Bacteria invade body — antigens on bacterial surface

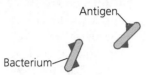

Antigen

Bacterium

2 Bacterial antigen recognised by the correct B-cell with the complementary receptor

Correct B-cell   Other B-cells with different receptor sites

3 The sensitised B-cell divides by mitosis to produce plasma cells and memory cells

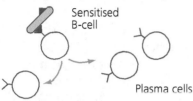

Sensitised B-cell

Plasma cells

Memory B-cell — remains in the body for a long time, providing immunity

4 Plasma cells produce antibodies, which destroy bacteria

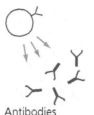

Antibodies   Antibodies destroy bacteria (e.g. causing cells to clump)

Figure 9 Antibody-mediated immunity

Synoptic links

Viruses

In AS Unit 1 you learned about the structure of viruses and that they function only within a host cell.

Cell-mediated immunity

Cloned T-cells form a variety of cells:

- **Killer (cytotoxic) T-cells** destroy infected cells directly by attaching to the antigens on the surface of the target (infected) cell and releasing the protein perforin, which produces pores within the cell-surface membrane, resulting in lysis and death of the target cell. Hydrolytic enzymes (called granzymes) are also secreted to aid in this destruction.
- **Helper T-cells** secrete cytokines (e.g. interferon, interleukin) which promote the activity of other cells: they stimulate B-cells to produce plasma cells and so increase antibody production; they activate killer T-cells; they activate macrophages and so stimulate phagocytosis.
- **Memory T-cells** do not act immediately but multiply very quickly if the antigen appears again, producing an even bigger crop of cloned T-cells, which results in the rapid destruction of any cells that present the antigen.
- **Suppressor T-cells** are sensitive to circulating cytokines and release their own cytokines after the successful elimination of the pathogen ('non-self' cell). This signals the deactivation of T-cells and/or B-cells and so prevents an excessive reaction.

Knowledge check 12

A collision between an appropriate B-cell and its antigen is less likely than the collision between a memory cell and the same antigen. Explain why.

Cytokines Cell-signalling molecules (peptides, proteins or glycoproteins) that aid cell-to-cell communication in immune responses and stimulate the movement of cells towards sites of infection and inflammation.

Interferon A type of cytokine important in defending the body against viruses. Interferons may be produced by genetically modified cells (a topic dealt with in A2 Unit 2).

Knowledge check 13

What is the difference between helper T-cells and killer T-cells?

The way in which cell-mediated immunity defends against viruses is illustrated in Figure 10.

1 Host cell infected with viruses presents viral antigens on its surface membrane

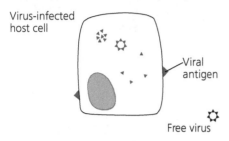

2 The viral antigen is recognised by the correct T-cell with the complementary receptor

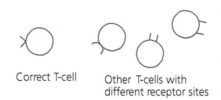

3 The sensitised T-cell divides by mitosis to produce different types of T-cell

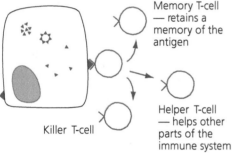

4 Killer T-cells recognise infected cells and destroy them before viruses reproduce

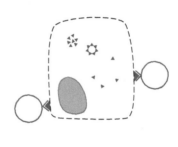

Figure 10 Cell-mediated immunity

The T and B systems complement each other in their action. For example, plasma (B) cell antibodies destroy 'free' pathogens but cannot deal with pathogens that are inside host cells; killer T-cells destroy those infected cells. Many bacteria damage tissue without entering cells, and so these, and any free viruses in the plasma, are dealt with by the B system; all viruses and a few bacteria invade host cells, and such infected cells are dealt with by the T system. T-helper cells enhance other parts of the immune system.

Killer T-cells destroy not only infected cells but also:

- **cancer (tumour) cells** within the host, since these present abnormal antigens on their cell-surface membranes
- **cells of transplanted tissue**, since cells from another person — unless that person is an identical twin — will have 'non-self' antigens

Active and passive immunity

Immunity may be acquired either actively or passively.

Passive immunity

Passive immunity occurs when an individual receives antibodies from another source. This can happen naturally or artificially.

Knowledge check 14

HIV (human immunodeficiency virus) invades and destroys helper T-cells and may lead to AIDS (acquired immune deficiency syndrome). Explain why people with full-blown AIDS suffer infections that would normally be held in check (called **opportunistic infections**).

- **Natural passive immunity** occurs when antibodies pass naturally from mother to baby across the placenta and in the mother's breast milk (especially in colostrum, the first-formed milk).
- **Artificial passive immunity** occurs when antibodies are administered by injection. The antibodies may be obtained from a person recovering from infection, or from an animal (injected with toxin) such as a horse, or as monoclonal antibodies produced by genetically modified mouse cells.

Passive immunity is only temporary since antibodies are used up in antigen–antibody reactions (and naturally denature) and the recipient has no plasma cells to make more (since they remain in the donor). However, passive immunity will provide a baby with protection until the child develops its own immune system. Passive immunity is the only way to save someone bitten by a fatally venomous snake or spider — the victim is injected with a **serum** containing antibodies (antitoxins) against the toxin. It also becomes important during an outbreak of a new or virulent disease with no known treatment — anti-Ebola antibodies were obtained from recovering patients and used successfully to treat other patients infected with the Ebola virus. Figure 11 shows what happens to the level of antibodies in an individual when these are received passively.

> Serum Blood plasma without clotting factors, but containing antibodies (so more specifically referred to as **anti-serum**.)

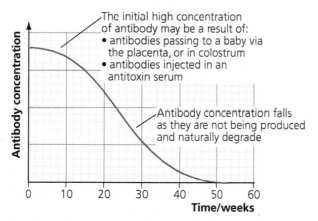

Figure 11 Changes in the concentration of antibodies during passive immunity

Knowledge check 15

Explain why bottle-fed babies are more likely to have infections than breast-fed babies.

Active immunity

Active immunity occurs when an antigen enters the body and stimulates the body's immune system to produce antibodies (and killer T-cells) and, most importantly, memory cells. Since memory cells can last a lifetime, active immunity provides long-term immunity. Active immunity happens naturally when you get a disease, or it may be artificially induced by vaccination.

Natural active immunity

Natural active immunity happens when a person is infected. On the first occasion, the person suffers the disease while B- and T-cells are activated and cloned — there is a time delay until antibodies are produced. In this **primary response**, while antibodies eventually destroy the pathogen, fewer and fewer B-cells are made and the concentration of antibodies falls again. If the same person is later infected with the same antigen (pathogen), the response is more rapid and a greater quantity of antibodies is produced. This **secondary response** is due to the action of memory cells:

■ The response is rapid because the memory cells have already been activated and cloning takes place immediately.
■ The high magnitude of the response is due to very large numbers of plasma cells being produced.

As a result, a person will not (normally) suffer the same infection twice. The primary and secondary responses are shown in Figure 12.

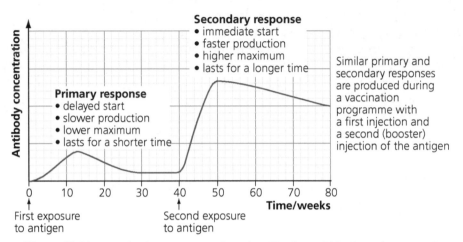

Figure 12 Changes in the concentration of antibodies within the primary and secondary responses during active immunity

Artificial active immunity: vaccination

Artificial active immunity involves vaccination (or immunisation). **Vaccination** was developed because some diseases are so serious that you really would not want to suffer their consequences even once. It involves injecting a person (or animal) with antigenic material that has been rendered harmless (e.g. by destroying the pathogen's nucleic acid) but leaves the surface antigens intact. As a result, the immune system manufactures antibodies and memory cells. The memory cells provide the long-term immunity. For some diseases, a vaccination programme involves a second injection of the vaccine, or **booster**. This further heightens the immune system — the level of antibodies produced by the two injections is similar to that produced as a result of two bouts of infection, as illustrated in Figure 12.

The different types of immunity are summarised in Table 3.

Table 3 Summary of different types of immunity

	Active: body receives antigens	Passive: body receives antibodies
Natural	Natural active: fighting infection	Natural passive: from mother via placenta or milk (colostrum)
Artificial	Artificial active: vaccination, involving injection of antigens or attenuated (weakened) pathogen	Artificial passive: injection of antibodies, e.g. antitoxins

Knowledge check 16

For B and T lymphocytes, give two differences and two similarities.

Knowledge check 17

How would the blood of someone who was immune to a disease be different from the blood of someone who was not immune to that disease?

Knowledge check 18

Some vaccines are given more than once, for example the MMR vaccine is given at the age of one year and then again at three years of age. Explain why.

Knowledge check 19

A person is vaccinated against TB. Is this active or passive immunity? Explain why.

Knowledge check 20

Vaccination cannot be used to treat a person once they have already been infected with a pathogen. Explain why.

The importance of vaccination in society

By giving long-term immunity to many diseases, vaccines have been important in preventing formerly common and dangerous epidemics. Through their use, smallpox has been eradicated worldwide, while measles, polio, rubella and tetanus are now uncommon in Britain and Ireland. Presently in the UK, vaccination offers protection from diphtheria, tetanus, whooping cough, polio, meningitis, measles, mumps, rubella and TB. This protection relies on a high proportion of the population being vaccinated, since chains of infection are likely to be disrupted. The greater the number of individuals who are immune, the smaller the probability that those who are not immune will come into contact with an infectious individual, a situation known as herd immunity.

Vaccination programmes over the past two centuries have caused infant mortality rates to plummet, allowed greater educational attainment, and improved the quality of family life and the vibrancy of the community in general. There are additional economic benefits: reduction in the medical costs of treating disease (which would otherwise be much greater than the cost of vaccination programmes) and, with fewer days off work, improved productivity.

Using antibodies to detect protein biomarkers

A monoclonal antibody can be produced to bind to a single, specific protein (antigen). It can be used in immunoassay, to detect the presence of a protein, acting as a biomarker, associated with a particular medical condition.

Examples of this use include:

1 **Diagnosis of prostate cancer:** if the prostate gland becomes cancerous, abnormally high levels of the protein PSA (prostate-specific antigen) may occur in the blood. This is tested for by using a specific monoclonal antibody, with an enzyme attached to allow the presence of the antibody to be detected. Such a test is called an ELISA (enzyme-linked immunosorbent assay) test (Figure 13).

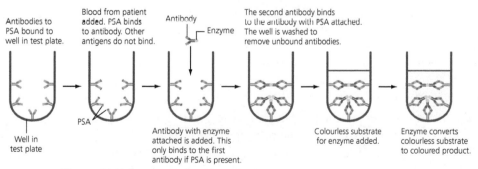

Antibodies to PSA bound to well in test plate.

Well in test plate

Blood from patient added. PSA binds to antibody. Other antigens do not bind.

PSA

Antibody
Enzyme

Antibody with enzyme attached is added. This only binds to the first antibody if PSA is present.

The second antibody binds to the antibody with PSA attached. The well is washed to remove unbound antibodies.

Colourless substrate for enzyme added.

Enzyme converts colourless substrate to coloured product.

Figure 13 Using monoclonal antibodies to test for prostate cancer

Herd immunity (or community immunity) Occurs when a large percentage of a population has become immune to an infection, so providing a measure of protection for individuals who are not.

Monoclonal antibody Antibody produced from a single clone of plasma cells (cultured from an isolated mouse B-cell in a laboratory). Thus, they are identical and are complementary to a single, specific antigen.

Immunoassay Test to measure the presence of a molecule through the use of an antibody.

Biomarker Naturally occurring molecule, often a protein, which can be detected and acts as an indicator of a disease or the effects of its treatment.

Knowledge check 21

Refer to Figure 13.
(a) Why do antigens other than PSA not bind to the first antibody?
(b) Why is the well washed at different stages in the test?
(c) Why does the second antibody have an enzyme attached to it?

2 **Detection of cytokines:** cytokines are small proteins, operating as molecular messengers between cells, released by inflamed tissue (as well as cells in the immune system). They can act as a biomarker of an inflammatory condition, which may be due to tissue injury (e.g. through cardiovascular disease) or bacterial infection. An ELISA test can be used to measure the level of cytokine in a blood sample (and so allow the detection of a medical problem, such as atherosclerosis or rheumatoid arthritis).

Organ transplant and 'non-self' antigens

Proteins (and glycoproteins) on the cell-surface membrane act as labels or markers and, in any one individual, are treated by the immune system as 'self'. However, since people differ genetically (except in the case of identical twins), different individuals will have different molecules on the surface of their cells. This presents a major problem when an organ or some tissue is transplanted from one individual to another. The recipient's body will tend to 'reject' the donated organ due to the activity of killer T-cells.

Successful organ transplants rely on the following:
- **Tissue typing**, where the compatibility of donor and recipient cell-surface molecules (markers) is first determined, then donor tissue is used for which there is an optimal match (i.e. most markers are the same). Tissue matching is more likely to occur between relatives (especially close ones) than between non-relatives.
- **Use of X-rays** to irradiate bone marrow and lymph tissues, so as to inhibit the production of lymphocytes and therefore slow down rejection. However, unpleasant side-effects occur and the patient is at increased risk of infection while the treatment is going on.
- **Immunosuppression** through the use of drugs, some of which inhibit DNA replication, cell division and the cloning of lymphocytes, and so delay the rejection of the graft. Again, problems may develop, including an increased susceptibility to infection.

Red blood cell antigens

Blood can be classified into types (groups) according to different markers on the surface of red blood cells. These markers act as antigens and affect the ability of red blood cells to provoke an immune response. Since the markers fall into distinct types they represent blood group polymorphisms (i.e. many distinct forms).

ABO group and blood transfusion

The **ABO blood system** is the most important blood-type system because of the presence of anti-A and anti-B antibodies in people who lack the corresponding antigens from birth (see Table 4).

Table 4 The antigens and antibodies of the ABO blood group system

Blood group	Antigens on red blood cells	Antibodies in plasma
A	A	anti-B
B	B	anti-A
AB	Both A and B	Neither anti-A nor anti-B
O	Neither A nor B	Both anti-A and anti-B

Anti-A and anti-B are agglutinins, i.e. if they encounter red blood cells with the complementary antigen they will cause them to stick together or agglutinate (clump; see Figure 14).

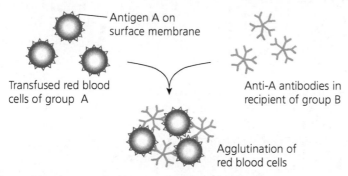

Figure 14 Donor blood group A will be agglutinated by anti-A antibody in the plasma of a recipient with blood group B

This effect is of great importance in **blood transfusion**: if donated blood were to agglutinate in the recipient's blood system, the clumps (note: not *clots*) of blood formed would block capillary networks and cause organ failure and death. The donated blood does not have to have the same blood type as the recipient. However, it is important that the donated blood does not have antigens that would be recognised by any of the antibodies present in the recipient. The 'safe' and 'unsafe' transfusions are shown in Table 5.

Table 5 Safe [✓] and unsafe [✗] transfusions: the recipient must not possess antibodies that will recognise antigens on the red blood cells of the donor

Blood group of transfused blood	Blood group of recipient			
	A	B	AB	O
A	✓	✗	✓	✗
B	✗	✓	✓	✗
AB	✗	✗	✓	✗
O	✓	✓	✓	✓

Blood of group O lacks the A and B antigens and is safe to use in all transfusions, so people of this group are called **universal donors**. Blood of group AB lacks both anti-A and anti-B antibodies, so people with this type can safely receive blood of any blood type and are called **universal recipients**.

Rhesus antigen and pregnancy

Other blood group systems exist. The **Rhesus system** depends on whether a marker protein is present. Most people have the rhesus factor (antigen) and are rhesus positive (Rh +ve); people who lack the factor are rhesus negative (Rh −ve). While Rh +ve people do not produce a complementary antibody, Rh −ve will produce antibodies against the rhesus factor (anti-D) if the antigen was to enter the body.

There are particular complications if a pregnant woman is Rh −ve and the baby is Rh +ve (fetal blood cells may be forced across the placenta as a result of severe uterine contractions during delivery). Generally, the first Rh +ve baby is born before the

Exam tip

Don't confuse the terms *clumping* and *clotting*. Clumping (or, more properly, agglutination) is the sticking together of red blood cells with antibodies acting as a molecular glue, while clotting (which you learned about in AS Unit 2) involves the formation of fibrin and only involves red blood cells in that they might incidentally become enmeshed in the network of fibres.

Rh −ve mother produces anti-D. However, any subsequent Rh +ve baby may receive anti-D from the mother (where memory cells rapidly produce plasma cells), with the result that the baby's blood agglutinates. One solution is to inject a Rh −ve mother with anti-D immediately after she has given birth to a Rh +ve baby. This injected anti-D will destroy fetal blood cells in the mother's body before the latter is sensitised to make its own antibodies.

Antibiotics and antimicrobials

By strict definition, antibiotic refers to substances produced by microorganisms that act against bacteria. Thus, antibiotics do not include antimicrobial substances that are synthetic (e.g. sulphonamides), or come from animals (e.g. lysozyme) or plants. In contrast, antimicrobials include all agents that act against all types of microorganisms — bacteria (antibacterial), viruses (antiviral), fungi (antifungal) or malaria (antimalarial).

Antibiotics and antibiotic resistance

Antibiotics are used to treat bacterial diseases — they are not effective against viruses. Penicillin prevents the production of the bacterial cell wall, so that when the cells absorb water they burst and die — penicillin is a bactericidal antibiotic. Tetracycline binds to bacterial ribosomes and inhibits protein synthesis — tetracycline is bacteriostatic, since it prevents bacteria reproducing rather than killing them.

The development of antibiotic resistance

While antibiotics have been hugely successful in treating bacterial disease such as tuberculosis (TB), many are now less effective as bacteria have developed **antibiotic resistance**. This happened because a few bacteria naturally possessed alleles of genes that coded for ways to prevent the effect of the antibiotic, e.g. bacteria resistant to penicillin possess enzymes which break down penicillin. With the widespread use of an antibiotic, those susceptible die while the resistant bacteria survive and reproduce so that the resistant strain predominates. (Alleles that confer resistance are often found on plasmids in bacteria — plasmids can be passed from one bacterium to another, and even between different species of bacteria.)

Controlling the spread of antibiotic resistance

To reduce the risk of antibiotic resistance developing, it is important that:

- antibiotics are used only when necessary — not for viral diseases or diseases that will be dealt with by the immune system, or given to livestock on a regular basis to increase growth rates
- a person prescribed antibiotics completes the course, as this increases the chances of eradicating all the disease-causing bacteria in the body
- hospitals and health centres take great care not to spread bacteria from one patient to another, by ensuring that strict hygiene regulations are followed (e.g. use of bactericidal hand gels)
- consideration is given to using more than one antibiotic to treat a disease, since it is unlikely that any one bacterium will possess two different resistance alleles
- some antibiotics are seldom used and kept as a 'last resort' when all others have failed

> **Exam tip**
>
> Fungi and bacteria are major decomposers (feeding on dead organic matter) and the production of antibiotics by some soil fungi (e.g. *Penicillium*) gives them a competitive advantage over bacteria living in the soil.

> **Knowledge check 22**
>
> Explain why neither penicillin nor tetracycline has any effect on viruses.

The consequences of antibiotic resistance

Nevertheless, some bacteria have become resistant to many antibiotics, notably methicillin-resistant *Staphylococcus aureus* (MRSA) and *Clostridium difficile* (*C. diff*). Both have caused deaths in hospital patients with suppressed immune systems, such as the elderly and those who have had organ transplants. The numbers of MRSA and *C. diff* cases have decreased in recent years as a result of specific strategies, such as implementing strict hygiene controls in hospitals.

Presently, patients undergoing routine surgery are given antibiotics to prevent infection. With the increased frequency of antibiotic-resistant bacteria, sepsis is becoming a common cause of death (44 000 people a year in the UK, more than from bowel, breast and prostate cancer combined!).

Discovering new sources of antibiotics

While no new antibiotic has entered clinical use since 1987, scientists are convinced that many environments harbour bacteria which produce antibiotics (to allow them to fight off other competing bacteria). A search of the ocean sediments revealed, in 2012, a bacterium (*Streptomyces* sp.) that produces anthracimycin, which seems to be effective against anthrax (*Bacillus anthracis*) and MRSA. It has a unique structure, though its mode of action is as yet unknown and it is still to be clinically tested.

Even more exciting is the development of a methodology for working with soil bacteria in situ (99% of which cannot be cultured in the laboratory). With this technique, scientists discovered a bacterium (*Eleftheria terrae*) from which a new antibiotic, teixobactin, was isolated in 2015. Teixobactin inhibits two types of lipid which act as precursors in bacterial cell wall synthesis and, since it has a dual action, it is unlikely that resistance to it will evolve. It has already been shown to be effective in the treatment of TB, septicaemia, *C. diff* and MRSA, though it has not yet been approved for medical use.

Apart from the soil and ocean sediments, scientists are looking at deserts, mountain tops and deep-sea hydrothermal vents as sources of exotic microbes. Other potential sources of new antimicrobials include honey, fish slime and crocodile blood.

Antimicrobials in plants

The global spread of drug-resistant pathogens has led to the search for alternative antimicrobials. To protect themselves, plants produce an array of compounds with an antimicrobial effect. Many have already been tested, including:

- fruit and vegetables, such as apple and spinach, which have been shown to contain antibacterial and antifungal compounds
- herbs and spices, such as caraway and thyme, with antibacterial, antifungal and antiviral compounds
- traditional medicines, such as St John's wort, containing antibacterials and antifungals, and quinine (from the bark of the Cinchona tree) with antimalarial properties

The many untested plants throughout the world may represent an untapped source of new antimicrobials. The potential discovery of new medicines in plants is an important reason for conserving the world's diversity.

Sepsis (also called septicaemia) The presence in tissues of harmful bacteria and their toxins, through infection of a wound.

Exam tip

You should be aware of the economic situation. Pharmaceutical companies are reluctant to invest large sums towards the discovery of new antibiotics since their function is to cure a disease and so their use is short term. Drugs for the treatment of cancer or cardiovascular disease, for example, are often given for life and so generate sufficient income to warrant research investment.

Knowledge check 23

Suggest why teixobactin resistance is unlikely to develop in infectious bacteria.

Exam tip

In practical work you may be asked to investigate the antimicrobial activity of fruit and leaf extracts. Discs of filter paper containing extracts are placed on an inoculated nutrient agar plate. Reduced growth of bacteria or fungi in the vicinity of the discs indicates antimicrobial activity.

Content Guidance

Practical work

Investigations with microorganisms involving aseptic techniques:
- the effect of different antibiotics/e-strips on bacteria, or
- the preparation of a steak plate to isolate single colonies

Investigation of the antimicrobial properties of plants.

Summary

- Resistance to infection by pathogenic organisms is possible because of barriers to entry, phagocytosis (by polymorphs and macrophages) and specific immune responses (antibody-mediated and cell-mediated).
- B lymphocytes are involved in antibody-mediated immunity while T lymphocytes are responsible for cell-mediated immunity.
- Antigens are specific molecules, found on the surface of invading pathogens (e.g. bacteria) or on infected cells, which trigger an immune response.
- In antibody-mediated immunity:
  - an inactive B lymphocyte with complementary receptors on its surface binds to a foreign (non-self) antigen on a microorganism or virus
  - the sensitised B lymphocyte divides to form plasma cells and memory B-cells
  - the plasma cells secrete antibodies that combine with the antigens on the pathogen and neutralise them (e.g. by agglutination)
  - the memory B-cells remain for many years but form plasma cells if the same antigen is detected in the future
- In cell-mediated immunity:
  - a T lymphocyte with complementary receptors binds to a foreign antigen displayed on an infected cell (or cell changed by cancer or cell transplanted from another individual)
  - the sensitised T lymphocyte divides to form helper T-cells, killer T-cells, memory T-cells and suppressor T-cells
  - helper T-cells produce chemicals that stimulate the B lymphocytes, killer T-cells and the phagocytes
  - killer T-cells release chemicals that punch holes in the cell-surface membrane, so destroying the infected cells
  - memory T-cells remain in the blood and respond rapidly if the antigen is met again
  - suppressor T-cells deactivate the immune response (of both B- and T-cells) when the infection is over
- During the primary response, to an initial infection, the host suffers disease symptoms. The secondary immune response, to a subsequent infection, is quicker and produces a higher concentration of antibodies so that the pathogen is dealt with before disease symptoms occur.
- Immunity can be stimulated artificially by vaccination. Since this is active in stimulating the production of plasma cells and memory cells, it provides long-term immunity. Vaccination has been important in preventing epidemics and pandemics.
- Antibodies can pass naturally into a baby (across the placenta and in the mother's milk) or artificially via an injection of serum containing antibodies (from a person recovering from infection) or of monoclonal antibodies (produced specifically for the antigen or toxin). Since this is passive, it is a short-term response only.
- A monoclonal antibody can be produced (in the laboratory) to bind to a specific protein. It can be used in immunoassay (or enzyme-linked immunosorbent assay, ELISA) to detect the presence of a protein, acting as a biomarker, associated with a particular medical condition.
- Transplanting tissue is possible only if the tissue is appropriately matched and medical techniques are used to suppress the immune system.
- Transfusion of blood is possible only if the recipient does not have antibodies complementary to the antigens on the red blood cells of the donor. Regarding the rhesus factor, there will be a problem if a Rh−ve mother has a second Rh+ve baby.
- Antibiotics are important for treating infections caused by bacteria. Antibiotic-resistant bacteria are becoming common, necessitating the need to find new antibiotics or antimicrobials.

Coordination and control

Coordination in flowering plants

Plants are able to detect environmental stimuli and respond to these through their growth and development.

Plant growth substances in stem elongation

Most plant responses are controlled by hormone-like chemical coordinators. These usually exert their effect by controlling plant growth, so are commonly called **plant growth substances**.

An outline of plant growth

During the growing season, cell division takes place at the tip (apex) of the plant in tissue called the **apical meristem**. These cells enlarge in a region below the tip called the **zone of elongation**; at certain times of the year the **internode** — the region between the nodes, or points at which leaves develop — also elongates (see Figure 15).

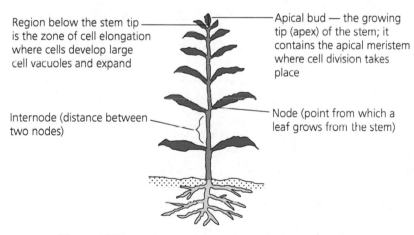

Region below the stem tip is the zone of cell elongation where cells develop large cell vacuoles and expand

Apical bud — the growing tip (apex) of the stem; it contains the apical meristem where cell division takes place

Internode (distance between two nodes)

Node (point from which a leaf grows from the stem)

Figure 15 The main growing regions of a flowering plant

The main effects of the major plant growth substances are shown in Table 6.

Table 6 The main effects of three plant growth substances

Plant growth substances	Site of production	Main plant growth effect
Auxins	Growing tip of stem (apical meristem)	Stimulate elongation of cells (in the zone of elongation)
Cytokinins	Actively dividing (meristematic) tissues	Promote cell division, especially in combination with auxins
Gibberellins	Apical buds and leaves	Stimulate elongation of internodal regions

Knowledge check 24

Suggest why only certain tissues in a plant will respond to a particular plant growth substance.

An outline of cell enlargement

The cells that divide by mitosis in the apical meristem are small, have thin cell walls and prominent nuclei, and lack large vacuoles. In the zone of elongation below, these newly formed cells develop small vacuoles and absorb water by osmosis, and eventually large, permanent vacuoles are formed. As the vacuoles absorb water, the cells increase in size — they enlarge (see Figure 16).

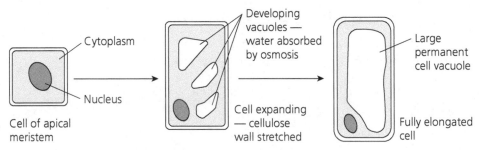

Figure 16 The process of cell enlargement

An outline of auxin action

- Auxins are produced in the cells of the apical meristem.
- They diffuse down the shoot towards the zone of elongation.
- They bind to specific receptors on the cell-surface membranes of the newly formed cells.
- This causes the membrane pumps to move hydrogen ions out into the cellulose cell wall.
- This acidification of the cell wall activates agents (such as elastins) which loosen the linkages between cellulose microfibrils, allowing slippage between them and making the wall more flexible.
- The cells absorb water by osmosis and the flexible cell walls allow the cells to expand as the extra water exerts increased hydrostatic pressure against them.
- The more auxin that is received in the zone of elongation, the more this effect allows cells to expand.

Phytochromes and the control of flowering

Some plants flower whenever they have grown sufficiently, irrespective of daylength (photoperiod). However, many flowering plants are sensitive to daylength and are classed according to the photoperiod in which they flower:

- **Long-day plants** (LDPs), e.g. poppy (*Papaver* sp.), flower only if the daylength exceeds a critical value.
- **Short-day plants** (SDPs), e.g. chrysanthemum, flower only if the daylength is less than a critical value.

The photoperiod is measured by **phytochrome pigments**, found in minute quantities in leaves. Phytochrome occurs in two interchangeable forms: P_{660} and P_{730}. P_{660} (P_R) maximally absorbs red light of wavelength 660 nm, while P_{730} (P_{FR}) maximally absorbs far-red light of wavelength 730 nm. Absorption of light by one form of phytochrome causes its rapid conversion to the other type. During darkness there is a slow conversion of P_{730} to P_{660}, which means P_{660} accumulates at night. Since daylight contains more red light than far-red light, P_{730} accumulates during the day. This information is summarised in Figure 17.

Figure 17 The interconversions of phytochromes (a) experimentally, and (b) in nature

P_{730} is the physiologically active form and can operate to either stimulate or inhibit flowering. The amount of P_{730} in the leaves is critical to flowering, and what influences that amount is the critical length of continuous (uninterrupted) darkness. The flowering requirements of LDPs and SDPs are summarised in Table 7.

Table 7 A summary of the flowering requirements of long-day and short-day plants

Class of plant	Required photoperiod for flowering	Relative level of P_{730}	Effect of P_{730}
Long-day plant (LDP)	Critically short period of darkness, i.e. shorter than a certain value	High level of P_{730}	Stimulatory
Short day-plant (SDP)	Critically long period of darkness, i.e. longer than a certain value	Low level of P_{730}	Inhibitory; removal of P_{730} required for flowering

The effect of different photoperiods, or light regimes, on the levels of the two forms of phytochrome and the subsequent effect on the flowering of long-day and short-day plants is illustrated in Table 8.

Table 8 The effect of different photoperiods on phytochromes and flowering in LDPs and SDPs

Photoperiod 0 ← hours → 24	Phytochrome response	Effect in LDP	Effect in SDP
Long day, Short night	P_{730} accumulates during the long day, and is not sufficiently removed during the short night — mostly P_{730}	Flowering	No flowering
Short day, Long night	P_{730} is removed during the long night and is not produced sufficiently during the short day — little P_{730}	No flowering	Flowering
Short day, Long night interrupted by short light period	During the night, the short light period converts P_{660} to P_{730} — enabling sufficient P_{730} to accumulate	Flowering	No flowering

Exam tip

Sunlight contains more light of wavelength 660 nm (red light) than 730 nm (far-red). Therefore, during daylight P_{660} (P_R) is converted to P_{730} (P_{FR}), which accumulates.

Exam tip

Historically, plants are categorised as short-day or long-day. This is unfortunate as it is the length of the dark period that is crucial and, indeed, it is the period of continuous darkness.

Exam tip

Know that P_{730} converts to P_{660} (slowly) in darkness and so is removed during a long night (short day) and know that P_{730} is the active form, either stimulating or inhibiting. You can then reason that an LDP flowers when the night is short (long day) so that sufficient P_{730} remains to promote flowering, while an SDP flowers when the night is long (short day) which is when P_{730} is removed, removing an inhibitory effect on flowering.

Manipulation of the photoperiod enables commercial growers to bring plants to the consumer at the time when they will command high prices, such as at Christmas. Thus chrysanthemums — short-day plants which usually flower in the autumn, as the days shorten — can be prevented from flowering by maintaining them under a long-day lighting regime. Understanding the phytochrome system allows further manipulation. As Table 8 illustrates, exposing the plants to a short period of light during the night can inhibit flowering. In this way flowering can be delayed for several months, until it is induced by a return to uninterrupted long nights — as when chrysanthemums can be readied for the market in time for Christmas.

The phytochrome system operates in leaves. Flowers develop in the terminal or axillary buds. These may be some distance away. This suggests the involvement of a chemical coordinator; however, as yet no such chemical has been isolated.

Knowledge check 25

A long-day plant normally flowers when the dark period becomes critically short. Explain what would happen if, before flowering was induced, the light period (daylength) was interrupted by a short period of far-red light.

Comparison of coordination in plants and animals

Coordination in both plants and animals involves receptors, a linking (communication) system and effectors.

- **Receptors** receive the **stimulus**: for example, in plants the phytochrome system in the leaf detects the photoperiod; in animals, the osmoreceptors in the hypothalamus are stimulated by changes in the water potential of the blood, while the retina of the eye is stimulated by light.
- **Effectors** bring about the **response**: for example, in plants a bud may develop into a flower, while cells elongate in the zone of elongation in response to auxin; in animals, muscle contracts — in the gut to bring about peristalsis, or in a limb to move one bone with respect to another.

Both plants and animals use chemicals to coordinate. For example, plants produce auxin in the tip and this moves down through the shoot. In animals, **hormones** are produced in one part of the organism and travel to other parts where they have their effects — for example, ADH is produced by neurosecretory cells in the hypothalamus (and stored in the posterior pituitary) and carried in the blood to the kidney, where it causes the collecting ducts to be permeable to water.

It takes time for chemicals to be produced and take effect, so there is a delay before the response takes place. The advantage is that the response can be maintained over a period of time.

However, animals need an additional system which is fast in linking receptors (e.g. in the retina of the eye) with effectors (e.g. muscles in the leg). This is because animals differ from plants in that they can move from one place to another (**locomotion**). This ability probably developed because animals have to search for food — unlike plants, which are autotrophic. Animals have a **nervous system** containing neurones, which transmit impulses very rapidly.

Summary

- Plant growth regulators include auxins, cytokinins and gibberellins; auxins stimulate the elongation of cells; cytokinins promote cell division; gibberellins stimulate elongation of the internodal regions.
- Auxins are produced in the shoot tip and diffuse down to the zone of elongation. They stimulate the pumping of hydrogen ions out of the cells and this increases the elasticity of the cell walls so that the cells elongate more during the process of enlargement (vacuolation).
- Plants have phytochrome pigments that detect the photoperiod. They are interchangeable: P_{660} is converted in the light (red light) to P_{730}, which slowly converts in the dark (rapidly by far-red light) back to P_{730} (the active form). The phytochrome system is involved in the control of flowering: in long-day plants an accumulation of P_{730} stimulates flowering; in short-day plants P_{730} inhibits flowering and so its removal (during a critically long night) results in flowering.
- In both plants and animals, chemicals are used to communicate between receptors (receiving stimuli) and effectors (effecting responses). However, plants lack a nervous system, a feature of animals that allows rapid communication and is important in coordinating movement.

Coordination in mammals

The mammalian nervous system can be divided into a **central nervous system** (CNS) and a **peripheral nervous system** (PNS). The CNS consists of the **brain** and **spinal cord**. It is responsible for integrating the activity of the nervous system in coordinating the functioning of all parts of the body. The peripheral nervous system consists of **cranial nerves** that are attached to the brain and **spinal nerves** that are attached to the spinal cord. These nerves connect receptors to the CNS and the CNS to effectors.

The **neurone** (nerve cell) is the functional unit of the nervous system. The human brain comprises tens of millions of neurones, each linked to other neurones. A **nerve** is a bundle of neurones.

Neurones and impulse propagation

The structure of neurones

Neurones link different cells and are able to conduct impulses between them. While variable in size and shape, all neurones have three parts:

- a cell body (**centron**), which contains the nucleus and other organelles, and has a number of cytoplasmic extensions
- extensions (called **dendrons** or, if they are very small and numerous, **dendrites**), which transmit impulses to the cell body
- extensions (**axons**), which transmit impulses away from the cell body and terminate in **synaptic bulbs** (or knobs) — axons may be more than 1 metre long

Two types of neurones of the PNS are shown in Figure 18: a sensory neurone, which conducts an impulse from a receptor to the CNS, and a motor neurone, which conducts an impulse from the CNS to an effector.

> **Knowledge check 26**
>
> Where would the cell body of a motor neurone be located?

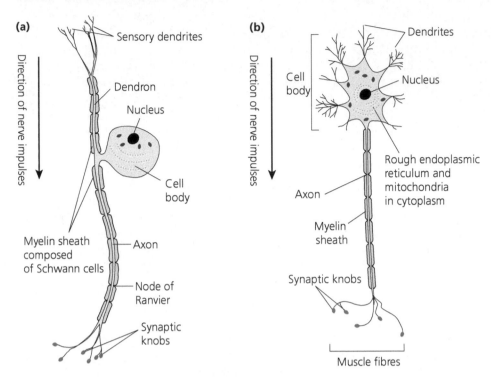

Figure 18 The structure of (a) a sensory neurone, and (b) a motor neurone

In mammals, many axons and dendrons are **myelinated** — covered with a **myelin sheath**. The myelin sheath is made up of many layers of cell membrane of **Schwann cells**, which wrap themselves round and round the axon. This insulates the axon, preventing ion movement between the axon and the tissue fluid around it. Each Schwann cell forms a sheath about 1 mm long. There are small spaces between Schwann cells, giving rise to gaps in the myelin sheath called **nodes of Ranvier**. At these nodes the axon is exposed.

The resting potential

All cells have a potential difference across their cell-surface membranes, due to differences in the distribution of ions. Neurones are special in having a particularly large potential difference. This is essentially due to a large excess of positively charged sodium ions (Na^+) on the outside. Since this potential difference occurs when the neurone is 'at rest', it is called the **resting potential**. The inside of the neurone is negative with respect to the outside and the magnitude of the resting potential is approximately −70 mV (millivolts).

The action potential

Neurones (and also muscle and receptor cells) are excitable — the potential difference can be reversed. When a neurone is stimulated, the cell-surface membrane allows sodium ions to diffuse in. This changes the potential difference across the membrane, making it less negative inside. If a critical potential difference (of about −55 mV), called the **threshold value**, is reached, then ions surge in and the neurone quickly becomes **depolarised**. Indeed, the cell will become positive on the inside, reaching a potential difference of about +40 mV. This reversal of the potential difference is called

Knowledge check 27

What is the name of the supporting cell that produces the myelin sheath?

Knowledge check 28

What is the main factor determining the resting potential of a neurone?

an **action potential**. An action potential does not vary in size and either occurs (if the threshold value is achieved) or does not, a phenomenon called the **all-or-nothing law**. The action potential is followed by a period when the membrane repolarises and recovers its resting potential. This recovery period is called the **refractory period**; during this time the membrane is unexcitable. These changes are shown in Figure 19.

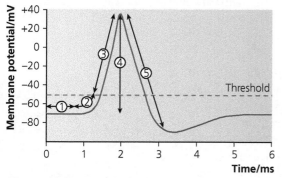

① Axon membrane at resting potential
② Depolarisation to threshold value needed for further depolarisation to occur
③ Further depolarisation leads to action potential
④ Magnitude of action potential: 40 mV – (–70mV) = 110 mV
⑤ Repolarisation of axon membrane during the refractory period when the membrane cannot be depolarised

Figure 19 Changes in the potential difference across the axon membrane as an action potential occurs and as it recovers its resting potential (note that these are changes over time at one point on the membrane)

Impulse propagation

An action potential that is generated in one part of a neurone is propagated rapidly along its dendron or axon. This happens because the depolarisation of one part of the membrane sets up **local circuits** with the areas either side of it. Local circuits occur as positive ions are attracted by neighbouring negative regions and flow in both directions. On one side, the membrane is still recovering its resting potential (repolarising), i.e. it is in its refractory period during which it cannot be stimulated. On the other excitable side, the local circuit triggers depolarisation and the formation of an action potential (see Figure 20). This process is repeated at each section of the membrane along the whole length of the neurone. In essence, then, an impulse is the transmission or propagation of depolarisations, or action potentials, along the neurone membrane.

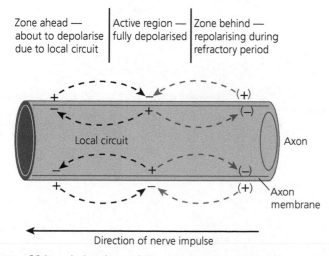

Figure 20 Local circuits and the propagation of an action potential

Exam tip

A membrane is said to be *polarised* if a potential difference is maintained across it with the inside being negative with respect to the outside. *Depolarisation* is a decrease and then temporary reversal of the potential difference so that the inside of the membrane becomes positive with respect to the outside. The membrane is actually still polarised but the polarity is reversed.

Exam tip

Local circuits are the movements of ions along the neurone. The flow of ions is caused by an increase in concentration at one point, which causes diffusion away from the region of higher concentration.

Knowledge check 29

Suggest why a refractory period is important for neurone function.

Exam tip

The nerve impulse is a wave of depolarisation. This is similar to the crowd in a sports stadium performing a 'Mexican wave'. At any one time only one section of the crowd is waving. The people do not move along, they stay in one place, but the wave travels along the crowd from one section to another. Similarly, one action potential does not actually travel along the neurone; a series of action potentials form in turn along the neurone.

In a myelinated neurone (such as a sensory or motor neurone in mammals), local circuits cannot be set up in the parts of the neurone insulated by the myelin sheath. Instead, the action potential 'jumps' from one node of Ranvier to the next. This greatly increases the speed at which it is propagated along the axon and is called **saltatory conduction**. Some non-myelinated axons have a large diameter and so a larger surface over which ions might be moved, thus the formation of action potentials occurs more rapidly and impulse transmission is speeded up.

Knowledge check 30

Suggest a disadvantage of the nodes of Ranvier being (a) closer together, and (b) further apart.

The synapse and synaptic transmission

At the end of an axon, the impulse will arrive at the synaptic bulbs which possess **synaptic vesicles** containing the neurotransmitter chemical. The synaptic bulbs come very close to their target cells (another neurone or muscle cell). There is a small gap between them called a **synaptic cleft**. The membrane of the neurone just before the cleft is called the pre-synaptic membrane and the one on the other side is the post-synaptic membrane. The whole structure is called a synapse (see Figure 21).

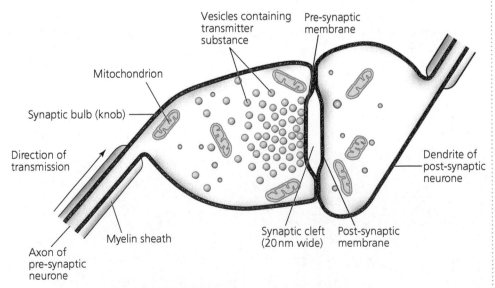

Figure 21 The structure of a synapse

Knowledge check 31

Explain the role of the mitochondria in the synaptic bulb.

The sequence of synaptic transmission is as follows:

- When an impulse arrives at the synaptic bulb, the membrane becomes permeable to calcium ions which diffuse into the bulb.
- Calcium ions stimulate movement of synaptic vesicles towards the pre-synaptic membrane.
- Vesicles fuse with the pre-synaptic membrane and release the neurotransmitter molecules (often **acetylcholine, ACh**) by exocytosis into the synaptic cleft.
- Neurotransmitter molecules diffuse across the synaptic cleft to the post-synaptic membrane — because the gap is only 20 nanometres (nm) wide, this takes only a fraction of a millisecond (ms) to occur.
- On the post-synaptic membrane, neurotransmitter molecules attach to specific receptors.
- This causes ion channels to open so that the potential difference in the post-synaptic membrane is altered. The magnitude of this change is dependent on the amount of transmitter released and so the number of receptors filled.
- When a molecule of ACh attaches to its receptor site, a sodium ion (Na^+) channel opens and the post-synaptic membrane becomes depolarised (less negative inside, as Na^+ ions diffuse in).
- Transmitter molecules on the receptors are inactivated. For example, ACh is hydrolysed by the enzyme **acetylcholinesterase** (abbreviated to cholinesterase) and the breakdown products, choline and ethanoic acid, released into the cleft. This breakdown of the neurotransmitter is important as it allows the resting potential to be re-established.
- The breakdown products diffuse across the cleft and are reabsorbed into the synaptic bulb where they are resynthesised into the neurotransmitter using energy in the form of ATP from mitochondria.

When a motor neurone synapses with a skeletal (voluntary) muscle fibre, there is a special kind of synapse called a **neuromuscular junction**.

Excitation and inhibition at synapses

While ACh is the main transmitter molecule (operating in the peripheral nervous system and so stimulating muscle contraction), there are other neurotransmitters, each with its specific protein receptor on the post-synaptic membrane. **Noradrenaline** is the neurotransmitter in the synaptic bulbs of the sympathetic nervous system (involved in the 'fight-or-flight' response). Both ACh and noradrenaline cause Na^+ ion channels to open, which results in depolarisation of the post-synaptic membrane and creates **excitatory post-synaptic potentials** (**EPSPs**). These EPSPs make the membrane less negatively charged and more likely to reach the threshold level to trigger an action potential, propagating an impulse in the post-synaptic cell.

However, the transmitter **gamma-aminobutyric acid (GABA)** is commonly found in the brain. When it attaches to its receptor sites, chloride (Cl^-) ion channels open, causing the post-synaptic membrane to be **hyperpolarised** (further polarised, as negative Cl^- ions diffuse in) so that **inhibitory post-synaptic potentials** (**IPSPs**) develop. These IPSPs make the membrane more

Knowledge check 32

Which mineral ion causes synaptic vesicles to fuse with the pre-synaptic membrane?

Exam tip

Neuromuscular junctions are always excitatory using ACh as the neurotransmitter.

negatively charged and less likely to reach the threshold level to trigger an action potential. Inhibitory synapses in the brain stop persistent worrying thoughts and so are important in controlling anxiety.

Even though synapses slow transmission slightly, there are distinct advantages. They:

- ensure that transmission is in one direction only, since synaptic knobs occur only at one end of a neurone, and the neurotransmitter is released only from the pre-synaptic side while receptors are located only on the post-synaptic membrane
- protect effectors (muscles and glands) from over-stimulation, since continuous transmission of action potentials exhausts the supply of transmitter substance, i.e. causes synapses to fatigue
- allow certain actions to be controlled through a combination of stimulation and inhibition, resulting in restraint as well as excitation
- integrate the activity of different neurones synapsing with a single post-synaptic neurone

The eye and the reception of light

The structure of the mammalian eye

The structure of the mammalian eye is shown in Figure 22. The main parts are described, along with their functions, in Table 9.

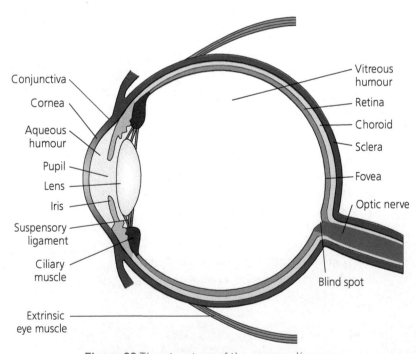

Figure 22 The structure of the mammalian eye

Knowledge check 33

A number of drugs or poisons exert their effect at synapses. Suggest the effect of (a) the South American arrow poison curare, which blocks the acetylcholine receptors on the post-synaptic membrane; (b) the nerve gas sarin, which inhibits the enzyme acetylcholinesterase.

Table 9 The main parts of the mammalian eye and their functions

Part of eye	Description	Function
Conjunctiva	Thin membrane lining the inside of the eyelids and covering the sclera (the 'white' region) near the front of the eye	Lubricates and protects tissues at the front of the eye; prevents foreign bodies entering the eye orbit
Sclera	Tough 'white' outer layer that encloses the eye, except at the front	Protects the eyeball against mechanical damage; allows attachment of eye muscles
Cornea	Transparent front part of the eye continuous with the sclera	Allows the passage of light while refracting (bending) it
Iris	The coloured part of the eye — it contains circular and radial muscles	Controls the size of the pupil to adjust the amount of light entering the eye
Pupil	A gap within the iris — appears black (light goes in and does not come out)	Allows light to enter the eye — its size can be varied
Aqueous humour	Transparent watery fluid filling the front part of the eye	Maintains the shape of the front chamber of the eye
Lens	Transparent and elastic ovoid (biconvex) structure held in place behind the cornea	Changes shape to adjust the focusing (accommodation) of light onto the retina
Ciliary body	Structure which supports the lens and contains circular muscle	Contraction or relaxation of the circular muscle controls the shape of the lens
Suspensory ligaments	Strong ligaments which connect the ciliary body to the lens	Transfers tension in the wall of the eyeball to the lens to make the lens thinner
Vitreous humour	Transparent, jelly-like material filling the rear part of the eyeball	Maintains the shape of the rear part of the eyeball and supports the lens
Retina	Inner layer of the wall of the eyeball containing the light-sensitive cells	The light-sensitive cells (rods and cones) initiate impulses in associated neurones when appropriately stimulated
Fovea	The region at the back of the retina that is rich in cones	A region with high visual acuity that allows colour vision
Choroid	A layer of pigmented cells at the back of the eyeball behind the retina	Contains blood vessels which supply the retina; prevents reflection of light within the eyeball
Optic nerve	A bundle of sensory nerve fibres which leaves from the back of the eye	Transmits impulses from the retina to the optic centre at the back of the brain
Blind spot	Region where optic nerve leaves inside of eyeball and so contains no light-sensitive cells	A region which, if light strikes it, is not sensitive to light

> **Knowledge check 34**
>
> What is the function of the cornea?

The iris and the control of pupil diameter

In bright light, the **circular muscles** contract and the **radial muscles** relax, making the iris widen (dilate) and therefore making the pupil get narrower. The pupil is small (constricted) to limit the amount of light passing through, as very bright light can damage rods and cones.

In dim light, the opposite takes place: circular muscles relax and radial muscles contract, widening (dilating) the pupil. The pupil is large to allow more light to reach the retina and so provide maximal stimulation of the light-sensitive cells there.

> **Exam tip**
>
> Don't try to just memorise how the iris functions, make sense of it. In bright light a small pupil is required (to prevent photoreceptor damage) so the iris is pulled towards the centre by contraction of circular muscle. In dim light a big pupil is required (to sufficiently stimulate photoreceptors) so the iris is pulled away from the centre by contraction of radial muscle.

The lens and the accommodation (focusing) of light

The cornea refracts light, but it is the lens that adjusts the refraction of light to a single point on the retina. The ability to adjust focusing is called **accommodation**. This occurs by means of adjustments to the shape of the lens.

The wall of the eyeball is under pressure, simply by being filled with fluid. If the **ciliary muscles** are relaxed, this pressure is transferred via the suspensory ligaments to the lens, pulling it into a thin shape. This means that the lens does not converge the light as much, which is what is required to accommodate light onto the retina from a *far object*.

To accommodate light onto the retina from a *near object*, the circular muscles in the ciliary body contract. This closes the aperture around the lens and releases any tension from the eyeball. The lens, being elastic, adopts a fatter shape, which means that it refracts the light more, so accommodating light onto the retina.

> **Exam tip**
>
> For focusing onto a far object, the lens is pulled into a thin shape. Don't say that this is due to *contraction* of the suspensory ligaments. The suspensory ligaments simply transfer the tension in the wall of the eye through to the lens.

The structure of the retina

The retina of the human eye contains two types of cells which are receptors to light: **rods**, which are sensitive to dim light; and **cones**, which respond in bright light only, but are able to discriminate fine detail and distinguish different wavelengths (colours) of light. Cones are concentrated at the fovea (yellow spot), at the centre of the retina (see Figure 22), while rods are mainly found around the periphery of the retina. The rods and cones synapse with bipolar neurones which, in turn, synapse with neurones of the optic nerve. Significantly, many rods synapse with each bipolar neurone and many bipolar cells connect with each neurone of the optic nerve. This is called **retinal convergence**. Cones, meanwhile, generally synapse with a single bipolar neurone and a single neurone of the optic nerve. Note that the light passes through the neurones before reaching the outer segments of the rods and cones. The consequence of this is that for the neurones to leave the eye they must pass through the layer of photoreceptors, creating an area devoid of receptors — the **blind spot**. The retina is shown in Figure 23.

> **Knowledge check 35**
>
> What is the shape of the lens when focusing on a near object?

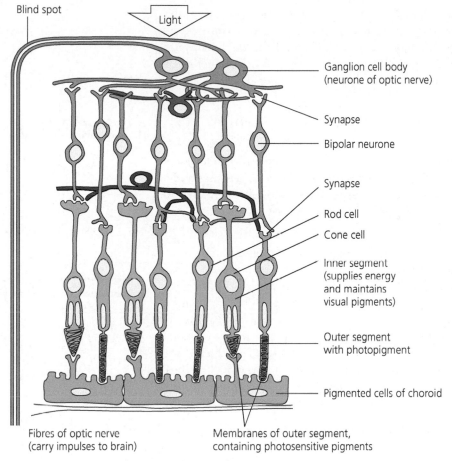

Blind spot

Light

Ganglion cell body
(neurone of optic nerve)

Synapse

Bipolar neurone

Synapse

Rod cell

Cone cell

Inner segment
(supplies energy
and maintains
visual pigments)

Outer segment
with photopigment

Pigmented cells of choroid

Fibres of optic nerve
(carry impulses to brain)

Membranes of outer segment,
containing photosensitive pigments

Figure 23 The structure of the retina

Exam tip

Examine Figure 23 and you will see that retinal convergence occurs at two levels: a number of rod cells converge on one bipolar neurone, and a number of bipolar neurones synapse with a single neurone of the optic nerve.

Rods and vision in dim light

The outer segment of a rod cell contains many membranes, all stacked up in parallel to each other. These membranes contain a pigment called **rhodopsin**, which is composed of a protein (opsin) and a light-absorbing compound derived from vitamin A, called retinal. When light strikes rhodopsin, changes occur and rhodopsin breaks down into retinal and opsin. This results in a change in the membrane potential of the rod cell and creates a **generator potential**. This causes a change in the membrane potential of the neighbouring bipolar neurone, with which the rod cell synapses. The bipolar cell releases transmitter substance into its synapse with a neurone of the optic nerve. If sufficient transmitter substance is released, an action potential is generated in the neurone of the optic nerve and transmitted to the visual centre at the back of the brain. Rod cells are sensitive, partly because rhodopsin absorbs light readily and is more easily broken down and, to a greater extent, because retinal convergence enables the input of many rods to be added together. This **summation** occurs because generator potentials have an additive effect in stimulating bipolar neurones while transmitter substance from bipolar neurones is added to reach the threshold needed to initiate an action potential in the neurone of the optic nerve. However, the consequence of retinal convergence is that the brain cannot distinguish which rod of a group sharing the same optic nerve fibre has been stimulated, i.e. there is decreased visual acuity and rods lack the ability to discriminate detail.

Exam tip

If asked to explain summation, don't say that impulses are added together — this is simply wrong. What you should say is that, with low-intensity stimuli, generator potentials have an additive effect and that there is enough transmitter released to reach the threshold for an action potential to be generated (and an impulse propagated) in a neurone of the optic nerve.

Rhodopsin is continuously re-synthesised within the rods. This re-synthesis requires ATP from the mitochondria. The rate of re-synthesis is sufficient for the rods to continue functioning in dim light. However, in bright light the rhodopsin is almost entirely bleached. It takes about 30 minutes in complete darkness for all the rhodopsin to be re-formed and so for the rods to become functional, a phenomenon called **dark adaptation**.

Cones, visual acuity and colour vision in bright light

Light reception in cones is very similar to that in rods. The pigment involved here is **iodopsin**. Iodopsin is less readily broken down and cones produce a generator potential in bright light only. Furthermore, cones do not exhibit convergence and one cone associates with one neurone of the optic nerve. Each cone cell stimulated will generate an impulse to the brain. This gives high **visual acuity** — the brain is able to distinguish between points that are close together.

However, iodopsin exists in three different forms and there are three different types of cone, each with a different type of iodopsin. Each is sensitive to different wavelengths of light corresponding to the colours blue, green and red. This forms the basis of the **trichromatic theory of colour vision**. Pure red light will only break down the red iodopsin and only the red cones will fire impulses to the brain. This is interpreted by the brain as red. However, yellow light will break down some of the red iodopsin and some of the green iodopsin and so both red and green cones will fire impulses to the brain. This is interpreted by the brain as yellow. Thus colour perception in the brain depends on the relative proportions of the different types of cone that are firing impulses.

Binocular vision and visual fields

Some two-eyed animals, usually prey animals (e.g. rabbits), have their eyes positioned on opposite sides of their head so as to give the widest possible field of view. A wide visual field facilitates the detection of potential predators. Other two-eyed animals, usually predatory animals or primates (including humans), have eyes positioned on the front of their head. The use of both eyes to view an object (**binocular vision**), creating a single image, allows more accurate judgement of distance. It also allows **stereoscopic vision** as a result of an increase in depth perception, so that the brain can create a three-dimensional image.

> ### Practical work
> Examine prepared slides or photomicrographs of the mammalian eye:
> - Identify the conjunctiva, cornea, iris, pupil, ciliary body, suspensory ligaments, aqueous and vitreous humours, retina, choroid, sclera, blind spot, optic nerve, rods and cones.

Muscle and muscle contraction

Muscle is an 'excitable' tissue and is capable of contraction. There are three main types of muscle: **skeletal**, **cardiac** and **smooth**. These are compared in Table 10.

Knowledge check 36

Explain why objects that appear brightly coloured in a well-lit room appear in dim light to be various shades of grey.

Knowledge check 37

According to the trichromatic theory of colour vision, to which colours of light are the three different types of cone sensitive?

Knowledge check 38

Distinguish between binocular and stereoscopic vision.

Table 10 A comparison of cardiac, smooth and skeletal muscle

	Skeletal	Cardiac	Smooth
Appearance	Muscle fibres are multinucleate, with distinct striations (bands)	Cells are striated and branched, forming a linked network; intercalated discs between cells	Spindle-shaped cells with a single nucleus and no striations
Distribution	Attached by tendons to bones	Found only in the wall of the heart	Present in the iris and ciliary body of the eye, and in walls of tubular organs, e.g. gut, blood vessels and bladder
Function	Movement of parts of the body and locomotion; strong contractions but not long-lasting	Pumping of the heart maintains blood circulation; cardiac muscle will contract throughout life	Movement of materials within the body; peristalsis in the gut can continue for long periods

The structure of skeletal muscle

Skeletal muscle consists of bundles of muscle fibres. A muscle fibre (see Figure 24) is multinucleate, resulting from the fusion of many cells, so is relatively large, e.g. up to 100 μm in width and as long as 300 mm. The nuclei lie just beneath the **sarcolemma** (the surface membrane), out of the way of the packed **myofibrils**, each of which is surrounded in **sarcoplasmic reticulum** and joined transversely by **T-tubules**, and between which there are numerous mitochondria. Each myofibril consists of overlapping arrays of thick and thin filaments composed, respectively, of the proteins **myosin** and **actin**.

Knowledge check 39

List the three types of muscle in order according to their tendency to fatigue, starting with the muscle type that fatigues least.

Exam tip

You need to be able to interpret myofibril (actin and myosin) arrangements in longitudinal and transverse sections, in both relaxed and contracted states.

Knowledge check 40

How does the appearance of each of the following alter when a muscle becomes contracted: I band, H zone, A band and sarcomere? Does it become longer, shorter or stay the same?

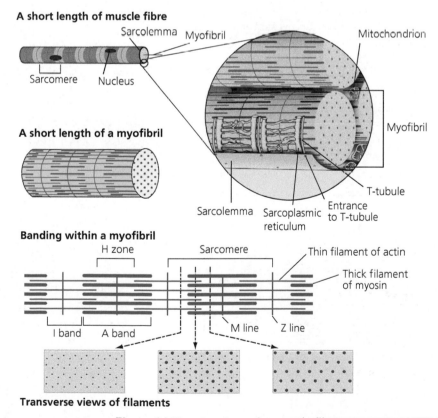

A short length of muscle fibre

Sarcolemma   Myofibril   Mitochondrion

Sarcomere   Nucleus   Myofibril

A short length of a myofibril

Sarcolemma   Sarcoplasmic reticulum   Entrance to T-tubule   T-tubule

Banding within a myofibril

H zone   Sarcomere   Thin filament of actin   Thick filament of myosin

I band   A band   M line   Z line

Transverse views of filaments

Figure 24 The structure of a muscle fibre

Muscle contraction

The shortening of myofibrils, according to **sliding filament theory**, causes muscle contraction. The sequence of myofibril shortening is as follows:

- An action potential arrives via a motor neurone at the synapse (neuromuscular junction) with the cell-surface membrane (sarcolemma) of the muscle fibre.
- Action potentials are propagated through the T-tubules and along the sarcoplasmic reticulum, causing calcium ions (Ca^{2+}), which are stored in the sarcoplasmic reticulum, to be released into the cytoplasm (sarcoplasm).
- Calcium ions cause ancillary proteins, which normally cover binding sites on the actin filaments, to be displaced and so uncover the binding sites.
- Heads of the myosin molecules next to the uncovered binding sites now attach to the actin filaments, forming acto-myosin 'bridges' between them.
- The myosin heads rotate or 'rock' back, pulling the thin actin filaments over the thick myosin filaments.
- ATP now binds with the myosin heads and the energy released from its hydrolysis (by ATPase) causes the myosin heads to detach from the actin filaments.
- The detached myosin heads regain the original position and attach to another exposed binding site on the actin filament, so that the cycle of attachment, rotation and detachment is repeated.
- This process continues as long as action potentials are propagated through the muscle fibre.

Figure 25 illustrates how the sliding of actin filaments over myosin filaments causes the myofibril, and the muscle fibre, to shorten.

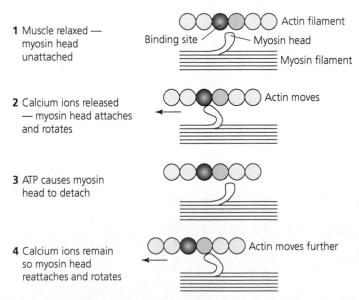

Figure 25 Movement of the myosin head causes the actin filament to slide over the myosin filament

Knowledge check 41

What is the role of
(a) calcium ions, and
(b) ATP in the sliding
filament theory of
muscle contraction?

Exam tip

To help you remember
the role of ATP in
muscle contraction,
you should note what
happens after death:
ATP production stops
and the cross-bridges
cannot break, so the
muscles stiffen in
a condition called
rigor mortis.

Exam tip

In a number of topics
you need to describe
a sequence of actions:
the transmission of a
nerve impulse; synaptic
transmission; the role
of the iris in controlling
the cone of light that
enters the eye; the
role of the lens in the
accommodation of light;
the role of rods and
cones in sensitivity and
visual acuity; muscle
contraction. You can
see animations of these
on certain websites.

Practical work

Examine prepared slides, photomicrographs and electron micrographs, as appropriate, of skeletal muscle, cardiac muscle and smooth muscle:
- Recognise the characteristic features of each.

Summary

Neurones and synapses:
- Neurones (sensory and motor) have a centron and long processes, dendrons or axons, which may have associated Schwann cells (forming a myelin sheath) separated by nodes of Ranvier.
- A non-conducting neurone has a resting potential across its membrane whereby it is negative on the inside.
- When stimulated, the neurone membrane is depolarised and becomes positive on the inside as an action potential develops.
- A nerve impulse consists of a wave of action potentials (depolarisations) moving along an axon.
- The nerve impulse travels faster in myelinated neurones as it jumps between the nodes of Ranvier (saltatory conduction). Impulses also travel faster along axons with larger diameters.
- Neurones are interconnected by synapses. Transmission across a synapse is achieved when a neurotransmitter, such as acetylcholine (or noradrenaline), is secreted into and diffuses across the synaptic cleft, and attaches to protein receptors on the post-synaptic membrane, which is depolarised as an excitatory post-synaptic potential (EPSP). If this reaches threshold, an action potential is generated and an impulse propagated in the post-synaptic neurone.
- Some synapses are inhibitory. In these, the neurotransmitter, such as GABA, causes the post-synaptic membrane to become more negative (an inhibitory post-synaptic membrane potential, IPSP, is generated) so that the generation of an action potential, and so impulse in the next neurone, is less likely.
- Synapses ensure that impulses are transmitted in one direction only and are vital in coordinating which neural pathways are selected.

The eye:
- The eye possesses the photoreceptor cells, rods and cones, within the retina.
- An iris controls the cone of light entering the eye: circular muscles contract to dilate the pupil; radial muscles contract to constrict the pupil.
- The lens accommodates light onto the retina: tension in the wall of the eyeball pulls the lens into a thin shape (for far objects); contraction of ciliary muscle relieves this tension and the lens becomes fatter (for near objects).
- Rods and cones contain photosensitive pigment, which is broken down by light to produce a change in potential difference. If this is large enough it generates an action potential and an impulse passes via the optic nerve to the brain.
- Rods and cones connect with the neurones of the optic nerve differently. Rods show convergence whereas cones do not. As a result, rods are more sensitive to dim light; cones have less sensitivity but produce an image with greater detail.
- In the trichromatic theory of colour vision, there are three types of cone (red, green and blue) sensitive to different wavelengths.

Muscle contraction:
- Mammals have three types of muscle: smooth, cardiac and skeletal (striated).
- Skeletal muscle consists of muscle fibres, which are multinucleate and possess many mitochondria, extensive sarcoplasmic reticulum (for storage of calcium ions) and infoldings of the cell-surface membrane called T-tubules.
- Muscle fibres comprise myofibrils consisting of actin (thin) and myosin (thick) filaments. Each myofibril has recognisable repeating units called sarcomeres, each of which contains overlapping filaments of actin and myosin.
- Contraction of the sarcomere (and so muscle) involves the following: the release of calcium ions unblocks binding sites on the actin filaments; myosin heads attach to the actin and the acto-myosin bridges rotate, causing filaments to slide over each other; ATP allows the myosin to detach and so the cycle of attachment–rotation–detachment is repeated.

■Ecosystems

Populations

A **population** is a group of organisms of the same species living in a particular habitat. As members of a population are of the same species, they will:

■ reproduce as long as there are available **resources** (any substance consumed by an organism, e.g. nutrients)

■ compete (**intra-specific competition**) if the resources are in limited supply

Synoptic links

Adaptation of organisms

In AS Unit 2 you learned about the adaptation of organisms to their environment, the ecological factors that have an influence on the distribution of organisms, and ecological techniques, such as the use of quadrats, randomly positioned, to estimate plant abundance.

Phases of population growth

For a new population starting in a particular area (e.g. in a laboratory culture or colonising a new habitat), four phases of growth are recognised. These are shown in Figure 26.

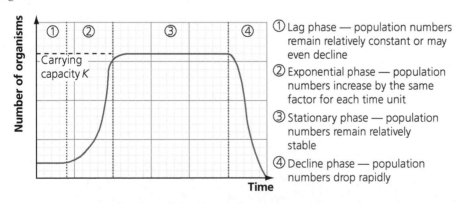

① Lag phase — population numbers remain relatively constant or may even decline

② Exponential phase — population numbers increase by the same factor for each time unit

③ Stationary phase — population numbers remain relatively stable

④ Decline phase — population numbers drop rapidly

Figure 26 The phases of population growth

> **Exam tip**
>
> The S-shaped curve is typical of any species that colonises new habitats. There is a period of slow growth as the species adapts to the habitat, followed by a period of rapid growth with little environmental resistance. The graph then levels off as the population reaches its carrying capacity. If one factor is in short supply, this can limit growth of the population, which then goes into decline.

Outline of the events in each phase

■ **Lag phase** — this is the time for nutrient assimilation, and/or egg production, and/or egg and larval development, or gestation period (in mammals).

■ **Exponential (log) phase** — reproduction creates new members of the population which add to the population's reproductive capacity (e.g. a bacterium divides to produce two bacteria, each of which produces two to give a total of four bacteria, and so on), and there is little competition since there are sufficient resources. The numbers increase by a value called the **intrinsic rate of natural increase** (designated **r**) and the population is illustrating its **biotic potential** (the reproductive capacity of a population under optimum environmental conditions).

■ **Stationary phase** — as population numbers increase, so resources become limiting and competition increases (and/or there is accumulation of waste, particularly in laboratory populations), i.e. there is **environmental resistance** (which prevents the population reaching its biotic potential). There is a decline in birth rate combined with an increase in death rate so that recruits (via births, and possibly immigration) equate with losses (via deaths, and possibly emigration), and the population size is determined by the resources available in the environment. The population reaches its **carrying capacity** (designated **K**), which is the maximum number that the environment can support.

■ **Decline phase** — the population has exhausted the resources and/or there is an accumulation of toxic waste so that the birth rate falls to zero and the death rate increases. Population numbers crash.

Renewable and non-renewable resources: the stationary and decline phases will depend on whether the resources are renewable (continually being replaced and made available to the organisms) or non-renewable (initially available but not replaced).

■ If resources are **renewable** (e.g. in woodland, trees continually fall and provide food for woodlice), the population will tend to remain in its stationary phase.

■ If resources are **non-renewable** (e.g. yeast grown in a laboratory batch culture), the population will have an exponential phase followed by rapid decline as the resources are consumed.

Exam tip

In nature, populations often do not have an obvious decline phase. For example, the fungi growing in a woodland soil will have the same large mycelium from year to year, with only seasonal fluctuations; this is because plant death and leaf fall renew the organic material for them to feed on. However, laboratory cultures of yeast will exhibit a decline phase and become extinct because they have used up the food source added initially.

Factors influencing populations may be grouped into two main categories:

■ **abiotic factors** — factors in the chemical and physical environment, e.g. carbon dioxide concentration, oxygen concentration, availability of mineral ions, water, light, temperature

■ **biotic factors** — the effects of other organisms of the same or different species

Intra-specific competition: competition between members of the same population will become more severe as the population increases and resources become limiting.

The influence of temperature: temperature is not a resource but will determine the metabolic rate in organisms and so the rate at which they develop. For example, it can be demonstrated that, in laboratory populations, the rate of increase (in the exponential phase) will rise at a higher temperature, but a higher temperature will not influence the size of the maximum population (in the stationary phase) — this will be determined by resources such as available nutrients. Similarly, a warm spring will produce rapid increases in insect populations (beneficial for the growth of insectivorous bird populations).

Knowledge check 42

For each of the following, compare the death rate and the reproduction rate: (a) lag phase, (b) exponential phase, (c) stationary phase.

Knowledge check 43

In a pond ecosystem, algae and pondweed compete for light and mineral ions. The algae are fed on by water fleas (*Daphnia*), which in turn are fed on by sticklebacks and *Hydra*. Identify the biotic and abiotic factors in the ecosystem described.

Knowledge check 44

Compare the influence of intra-specific competition during the exponential phase and the stationary phase of population growth.

Population dynamics

The number of individuals making up natural populations fluctuates over time: populations gain individuals through births or immigration (movements into the population) and lose individuals through deaths and emigration (movements out of the population). Population growth is determined by the equation:

population growth = births (B) − deaths (D) + immigration (I) − emigration (E)

A population in equilibrium, then, will have an equation: B + I = D + E.

Population dynamics: r-selected and K-selected species

Population dynamics of different species: two types of species are identified with respect to their reproductive strategy and the dynamics of their populations:

- The populations of the species increase rapidly as a resource becomes available and crash as the resource is used up, with repeated cycles of 'boom and bust'. Because of the prominence of the 'intrinsic rate of natural increase' (r), such species are called **r-selected species** or are said to have an **r-strategy**.
- The populations of the species remain at the carrying capacity of the environment (K). Such species are called **K-selected species** or are said to have a **K-strategy**.

Table 11 compares the features of r-selected and K-selected species. Most species have strategies between the two extremes.

Table 11 The features of r-selected and K-selected species

Feature	r-selected (r-strategist)	K-selected (K-strategist)
Length of life cycle	Short — quick to mature	Long — takes time before individuals become reproductively mature
Generation time	Short	Long
Numbers of offspring	Many	Few
Population density	Highly variable; often overshoots K, resulting in 'boom-and-bust' dynamics	Less variable; usually near K
Dispersal (ability to migrate)	High; species migrate readily and are able to re-colonise easily	Low; re-colonisation is uncommon
Competitive ability	Weak	Strong
Body size	Small	Large
Amount of parental care	Little	Considerable
Habitat	Unstable or disturbed	Stable and/or stressful

K-selected species are more prone to extinction as they cannot respond well to environmental disaster.

Knowledge check 45

What type of species has a growth curve characterised by occasional population explosions followed by population declines?

Practical work

Investigate the growth of a yeast population using a haemocytometer:
- components of a culture medium for a yeast population
- use of a haemocytometer to include counting cells over a grid of determined volume and the calculation of cell density

Estimate of the size of an animal population using a simple capture–mark–recapture technique:
- marking techniques
- assumptions made when using a capture–mark–recapture technique
- estimation of population size using the above technique

Exam tip

You may be asked to calculate estimates of population size: of a yeast population using a haemocytometer; of an animal population using the mark–release–recapture technique; of a plant population sampled using randomly placed quadrats.

Population interactions

Different populations within a habitat may affect each other's population growth. Table 12 shows three interactions: '+' indicates that there is a positive effect on one population, while '−' indicates that there is a negative effect and that the population would decrease.

Table 12 Three types of population interaction

Type of interaction	Effect on population growth	Comment
Mutualism	+/+	Both species gain; interaction may be necessary for both
Predation	+/−	The predator species gains, the prey species loses
Competition	−/−	Both species lose while interacting; the species most affected is eliminated from its niche

Knowledge check 46

Parasitism and predation are both +/− interactions. How do they differ?

Mutualism

If both species benefit from an interaction, their interaction is called **mutualism**. Nitrogen-fixing bacteria of the genus *Rhizobium* receive protection and nutrients within the root-nodule tissue of leguminous plants (such as clover, *Trifolium*) and provide the plant with nitrogen-containing compounds. Lichens are compound organisms consisting of highly modified fungi that harbour green algae among their **hyphae**. The fungi absorb water and nutrients and provide a supporting structure, while the algae carry out photosynthesis.

Predator–prey interactions

Predator–prey interactions change according to the relative numbers of prey and predator. An abundance of prey will mean that more of the predator population can be supported and so the predator population would grow. Large numbers of predators will reduce the prey population; few available prey would then not support the large numbers of predators, which would subsequently fall. The resulting predator–prey interaction will often produce oscillations, especially when the predator species hunts only one or a few prey species. Furthermore, the oscillations produced will show changes in the predator-species population which lag behind changes in the prey population (see Figure 27).

Knowledge check 47

The roots of many trees may grow in close association with the hyphae of particular fungi: the roots are able to absorb mineral ions released during decay by the fungi; the hyphae are able to absorb sugars from the plant. What type of interaction is being illustrated?

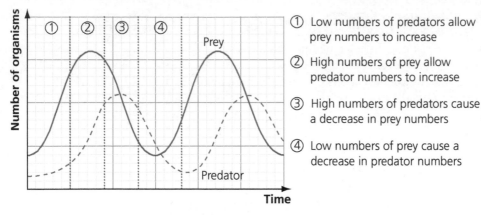

① Low numbers of predators allow prey numbers to increase

② High numbers of prey allow predator numbers to increase

③ High numbers of predators cause a decrease in prey numbers

④ Low numbers of prey cause a decrease in predator numbers

Figure 27 Predator–prey oscillations

Inter-specific competition

Inter-specific competition between the populations of two species occurs when they require a common resource which is in limited supply. The species may share the same niche or, at least, exhibit niche overlap. The characteristics of competition are:

- Both species do less well when competing for the resource.
- One species is eventually eliminated from the habitat.
- The winner may utilise the resource more efficiently and so be more successful, or it may have some feature the sole effect of which is to allow the winner to compete more effectively (e.g. the aquatic plant *Lemna gibba* has air sacs which allow it to float above aquatic algae, so *L. gibba* can absorb more of the available light).
- The outcome of competition may well be determined by the environmental conditions (e.g. the flour beetle *Tribolium castaneum* outcompetes *T. confusum* when conditions are warm and humid but not if it is cold and dry).

Figure 28 shows the population growth curves for two plant species when growing in isolation and when competing.

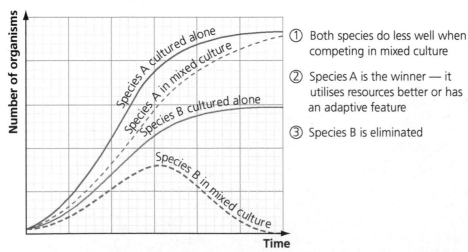

① Both species do less well when competing in mixed culture

② Species A is the winner — it utilises resources better or has an adaptive feature

③ Species B is eliminated

Figure 28 The population growth curves for two competing plant species when grown in pure and in mixed cultures

Knowledge check 48

Study Figure 27. Explain why (a) the rise and fall in the predator population always lags behind that in the prey population, and (b) the prey population is larger than the predator population (and indeed larger than shown in the graph).

Exam tip

Don't confuse the terms *intra*-specific and *inter*-specific competition. Intra-specific competition occurs between members of the same species; inter-specific competition occurs between members of different species.

Exam tip

You will need to be able to interpret data in graphs and tables that show changes in population numbers. For example, you should be able to analyse predator–prey and inter-specific competition curves, such as those in Figures 27 and 28. These skills should be practised. Past paper questions are a source of suitable material for practice.

The **competitive exclusion principle** is that two species cannot share the same niche without one species being eliminated. However, caution needs to be exercised: two species of grass might require the same mineral ions but obtain these at different levels in the soil because one has longer roots than the other; two species of plant would require light but avoid competition by growing at different times in the year.

Knowledge check 49

Why is it not possible for two different species to have identical ecological niches?

Biological control

A **pest** is any organism that competes with or adversely affects a population of plants or animals that is of economic importance to humans. **Biological control** aims to achieve permanent control of pest populations without the dangerous side-effects associated with chemical pesticides. A biological control method involves the introduction of a predator, parasite or pathogen (the **biological control agent**) to reduce the pest numbers to a point below the **economic damage threshold** (see Figure 29).

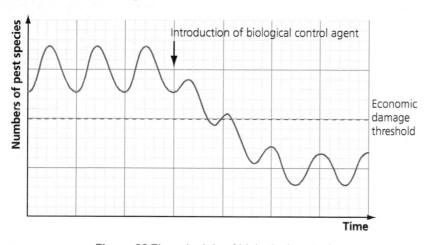

Figure 29 The principle of biological control

Advantages of using biological control agents over the use of pesticides include:

■ Biological control agents are more likely to target only the pest species.
■ Biological control agents do not have a negative effect on biodiversity.
■ The development of pests' resistance to biological control is unlikely.
■ Biological control agents reproduce and so are self-perpetuating over time.
■ Biological control is relatively cheap since, while the initial research may be costly, the predator or parasite does not have to be reintroduced (whereas pesticides must be re-applied often).

Knowledge check 50

List the properties of an organism used for the efficient biological control of a pest.

A biological control programme will require much research to ensure a successful outcome. For example, the control agent should target the pest species, it should not itself become a pest, it should survive and reproduce in its new environment, and no diseases should be imported with the control agent.

Biological control cannot cope with sudden surges in pest numbers and pesticides may need to be used in such circumstances.

Summary

- A population is all the individuals of a particular species in a particular habitat at a particular time.
- Population growth exhibits four stages:
  - Lag phase: the population is low and relatively stable as organisms adapt to their surroundings.
  - Exponential phase: the population increases rapidly because there are abundant (unlimited) resources.
  - Stationary phase: the population numbers are high and stable as births are matched by deaths. This is the carrying capacity of the environment.
  - Decline phase: overuse of resources or accumulation of waste products causes a decline in the population. If resources are renewed, the population may recover, otherwise it crashes to zero.
- The factors that limit population growth are known as environmental resistance. Such factors can be abiotic or biotic.
- Two types of population growth strategies are recognised:
  - r-selected species show rapid growth, exhausting resources quickly, and move on; a 'boom-and-bust' strategy.
  - K-selected species are more conservative — populations grow slowly though numbers remain relatively stable at the carrying capacity, unless disturbed.
- Population numbers are influenced by a number of biotic factors:
  - mutualism, where populations of different species benefit from the association
  - predator–prey, where one species (the predator) preys on another; oscillating population curves may result, with predator numbers lagging behind those of the prey
  - inter-specific competition, where two species compete for a common resource — the outcome is the removal of one species by the other
- Biological control is the use of predators to limit the numbers of a pest species. The control species should be highly specific (to the pest), long-term (remain in the ecosystem) and should not contaminate the environment (not itself become a pest).

Communities

A **community** consists of all the populations of organisms in a particular habitat. An **ecosystem** consists of a habitat and its associated community, so that it includes both biotic and abiotic components, and the interactions within and between them.

Community development

Ecosystems change constantly, with new species entering the community and others being lost. **Succession** is the term used to describe the progressive change in the species composition of a community over a period of time. There are two kinds of succession: primary and secondary.

Primary succession

A **primary succession** takes place in a previously uncolonised substrate. Examples include volcanoes erupting and depositing lava; a glacier retreating and depositing rock; landslides exposing rocks; sand being piled into dunes by sea or wind; lakes being created by subsiding land.

> **Exam tip**
>
> An examiner will expect you to define precisely, and distinguish between, the terms *population*, *community*, *ecosystem*, *habitat* and *ecological niche*. Develop your understanding of biological terms by maintaining a list of definitions.

A small number of **pioneer** plants, which are usually r-strategists specialised in dispersal to and colonisation of exposed areas, dominates new communities. This simple community modifies the abiotic environment (causes it to change). This results in a change in the community, which further alters the abiotic environment, and so on. Each successive community makes the environment more favourable for the establishment of new species. The process continues through a number of stages (called **seres**) until a final stage, the **climax community**, is formed. The climax community is a relatively stable end-stage and is in dynamic equilibrium with its environment. If the composition of the climax community is determined by the climate, it is called a **climatic climax**. In most of Britain and Ireland the climatic climax is deciduous forest. Some successions do not reach a potential climax because of interference by a biotic factor, such as grazing by deer, and the final community is called a **biotic climax**.

An example of succession from bare rock is illustrated in Figure 30. Weathering of the rock and the production of organic material by the pioneer species lead to the beginnings of soil formation. Further succession results in a deeper and more nutrient-rich soil.

Knowledge check 51

How might pioneer species make the next stage of succession possible?

Exam tip

If you are asked to define 'climatic climax community', you must attend to both *climatic* and *climax*. So a climatic climax community is a relatively stable community at the end-stage of succession in which the community composition is determined by aspects of the climate.

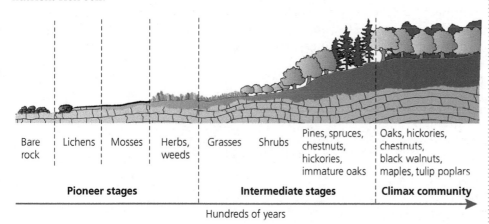

Figure 30 Ecological succession from bare rock in a North American community

As succession proceeds, the following trends are evident (see Figure 31):

- The soil develops — it increases in depth and in the proportion of organic content (humus).
- The height and biomass of the vegetation increase.
- Changes in height and density of vegetation provide a greater variety of microhabitats and ecological niches.
- The increasingly complex plant community and wider variety of niches support a greater number of animal species.
- Species diversity increases, from simple communities of early succession to complex communities of late succession.
- The number of food chains increases and more complex food webs develop.
- The community becomes more stable and becomes dominated by long-lived plants (K-strategists).

Knowledge check 52

During the succession shown in Figure 30, shrubs out-compete grasses and replace them. What factor are the grasses and shrubs likely to be competing for?

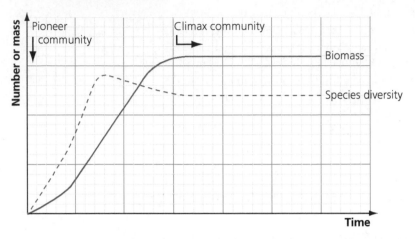

Figure 31 Some general changes in vegetation during succession

Succession takes place over long periods, e.g. hundreds of years. It is possible to 'view' succession in a sand-dune system where the zonation of vegetation over 'space' reflects changes over time. The dunes furthest from the sea were those formed first, i.e. they are the oldest; those nearest the shore were formed last, i.e. they are the youngest — see Figure 32, which also shows changes in abiotic conditions. You can see how this succession complies with the general theme (see Figure 31).

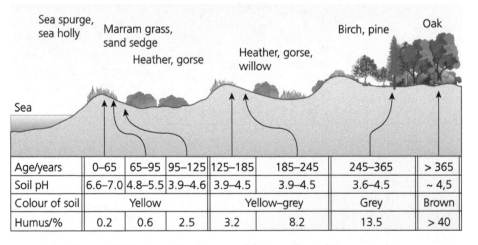

Age/years	0–65	65–95	95–125	125–185	185–245	245–365	> 365
Soil pH	6.6–7.0	4.8–5.5	3.9–4.6	3.9–4.5	3.9–4.5	3.6–4.5	~ 4,5
Colour of soil	Yellow			Yellow–grey		Grey	Brown
Humus/%	0.2	0.6	2.5	3.2	8.2	13.5	> 40

Figure 32 The zonation in a sand dune reflects its succession

Secondary succession

Secondary succession occurs after an existing ecosystem has been disturbed, e.g. by fire or flooding. Secondary succession happens more quickly than primary succession because the former type of event does not involve loss of soil and because some vegetative remains from the previous community are present. Table 13 shows the succession that would take place in an abandoned ploughed field.

Exam tip

You should be clear about the difference between *succession* and *zonation*. Succession refers to the change in the composition of a community with time. Zonation is where different communities are present at the same time, distributed as bands or zones along an environmental gradient.

Knowledge check 53

What is the difference between primary and secondary succession?

Table 13 Secondary succession in an abandoned ploughed field

Pioneer	Successive seres			Climax
Annual plants (live for only one year but produce vast numbers of easily dispersed seed)	Grasses and low-growing perennials (plants which live for many years)	Shrubs and small trees	Young broadleaved woodland trees	Mature woodland, mainly oak

Summary

- A community is all the organisms present in a particular habitat at a particular time, while an ecosystem is a community of organisms in which the organisms interact with each other and with the abiotic environment.
- Succession is the progressive change in the composition of a community over a period of time. The first organisms to colonise a habitat are pioneers. Colonising organisms bring about changes that allow other organisms to colonise the habitat. A series of additions and replacement of species occurs. Each distinct community in a succession is called a seral stage and the final stable community is called the climax.

- If the area has never been colonised before (e.g. a newly formed volcanic island), the process is called primary succession. If the area was previously covered with vegetation (e.g. a forest site that suffered a severe fire), secondary succession occurs.
- There are different types of climax community: the composition of a climatic climax is dependent on climate, such as temperature and rainfall; the composition of a biotic climax community is influenced by biotic factors, such as grazing.

Ecological energetics

Feeding relationships

Feeding involves the transfer of energy and materials from one organism to another. The sequence of organisms, with each being a source of food for the next, is called a **food chain**. A **grazing food chain** starts with living plants, fed on by herbivores, in turn fed on by carnivores, and so on, e.g. grass → rabbit → buzzard. A **detritus food chain** starts with dead organic matter (dead plants and animals) fed on by detritivores and decomposers. For example, dead grass → woodlice (detritivores) or fungi (decomposers). Since an organism usually feeds on several types of organism and in turn is fed on by more than one type, the result is a **food web**. For example, rabbits feed on herbaceous plants other than grass, while foxes and stoats, as well as buzzards, feed on rabbits.

Within a food chain or food web each organism occupies its own feeding position or **trophic level**:

- **Producers** — the autotrophic plants produce food through photosynthesis and ultimately support all other levels.
- **Consumers** — primary consumers (herbivores) feed on producers, secondary consumers (carnivores) feed on primary consumers, tertiary consumers ('top' consumers) feed on secondary consumers, and so on.
- **Decomposers** — decomposers (most bacteria and fungi) and detritivores (e.g. earthworms and woodlice) feed on the accumulated debris of dead organic matter (detritus).

Knowledge check 54

In a semi-desert ecosystem, seeds are eaten by mice, which in turn are eaten by snakes. Hawks feed on mice and snakes. What can you infer about the trophic level(s) occupied by hawks in the ecosystem?

Changes in any population in a food web can influence other populations. For example, in the marine ecosystem shown in Figure 33, a decrease in the number of crabeater seals will have a number of effects: it may mean that leopard seals feed more regularly on emperor penguins, which might decrease in numbers; it might mean more fish for Adélie penguins, which might increase in numbers.

Food chains commonly have four links and rarely more than five. The reason for this is that not all the energy of one trophic level is available to the next. There is an 'inefficiency' of energy transfer and at the fifth trophic level there is too little energy to support a further level.

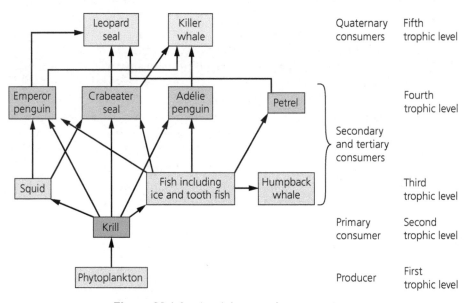

Figure 33 A food web in a marine ecosystem

Pyramid relationships

Food chains and food webs illustrate the types of organism only. Ecological pyramids show the quantitative relationship of the trophic levels. There are three forms:

- pyramid of numbers
- pyramid of biomass
- pyramid of productivity/energy

each with advantages and disadvantages.

A **pyramid of numbers** represents the total numbers of the organisms at each trophic level in an ecosystem. The data for constructing a pyramid of numbers are readily obtained: simply count the organisms in each trophic level within a specified area. However, there are disadvantages:

- With some organisms, it is difficult to determine what represents a single individual, e.g. with buttercups producing runners so that several are connected.
- The approach does not take into account the size of an organism, e.g. an oak tree has more producer material than one buttercup, while a single rabbit will have lots of mites.

While pyramid relationships are frequently illustrated, inverted pyramids are not uncommon (see Figure 34).

Knowledge check 55

Explain how, in a woodland ecosystem, a few producers may support a large number of consumers (resulting in an inverted pyramid).

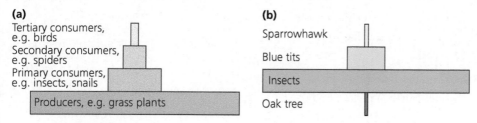

Figure 34 Pyramid of numbers for (a) grassland, and (b) an oak tree

A **pyramid of biomass** represents the total biomass of the organisms at each trophic level in an ecosystem. The data for constructing a pyramid of biomass are obtained by weighing the dry mass of the organisms at the different trophic levels. While this requires more effort than for a pyramid of numbers, it is more representative of all the material in each trophic level. Still, the information represents the **standing crop**, i.e. what is present at one moment in time and not what is being produced over time. As a result, inverted pyramids are possible, especially for aquatic ecosystems where the amount of the phytoplankton (producers) at any one moment in time may seem relatively small but is being generated exceedingly quickly and so supports a larger biomass of zooplankton (consumer level; see Figure 35).

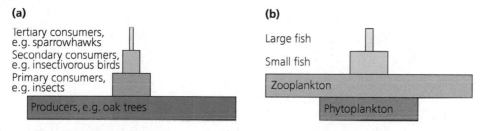

Figure 35 Pyramids of biomass for (a) woodland, and (b) a marine ecosystem

A **pyramid of productivity** (also called a **pyramid of energy**) represents the energy value of new material produced at each trophic level over time. It is more difficult to obtain the data simply because measurements need to be made over a time period — measurement may be presented as, for example, $kJ\,m^{-2}\,y^{-1}$ (kilojoules per square metre per year). However, comparing the *rate* at which the organisms at each trophic level produce new material will always produce the meaningful pyramid relationship (see Figure 36).

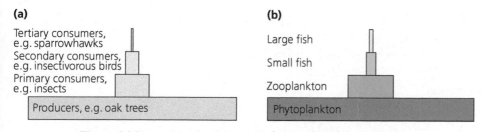

Figure 36 Pyramids of productivity (energy) for (a) woodland, and (b) a marine ecosystem

Knowledge check 56

Estimating the standing crop is relatively simple: all the plant material from a measured area (e.g. $1\,m^2$) is collected and dried (at 105°C) until it reaches a constant mass. Suggest why (a) dry mass is measured rather than wet mass, (b) the plant material is not heated above 105°C.

Knowledge check 57

In aquatic ecosystems, the first trophic level (phytoplankton) may have a lower biomass than the second level (zooplankton). Explain why.

Exam tip

You need to be able to analyse pyramid relationships and explain the limitations of pyramids of numbers and biomass. Further, you need to be able to interpret pyramid relationships for different ecosystems.

Productivity is the energy entering a trophic level that remains as energy in biomass:

- for a given area of the ecosystem (often a square metre)
- in a given period of time (often a year)

So productivity is the *rate* of biomass production (units, for example, $kg\,m^{-2}y^{-1}$) or of energy accumulation (units, for example, $kJ\,m^{-2}y^{-1}$).

The inefficiency of energy transfer in ecosystems

There are a number of reasons for the inefficiency of energy transfer and the precise nature of each of these depends on the trophic level.

Reasons for the low percentage of solar radiation absorbed by plants in photosynthesis: only a small fraction of the total amount of solar energy arriving at the Earth is used by the producers (i.e. in photosynthesis). At the equator during the day, about 1.4 kJ of energy every second reaches the upper atmosphere from the sun over every square metre. Most of this solar energy never reaches the ground since over 99% is:

- reflected back into space by clouds and dust, or
- absorbed by the atmosphere and re-radiated

Of the solar energy that reaches the ground, most will not be used since:

- most will miss the leaves altogether, with less than 0.1% actually reaching the surface of the leaves

When light from the sun does hit leaves, a mere 0.5–1% of the incident energy is actually used by a plant in photosynthesis since:

- some of the energy is reflected by the leaf, or
- some is transmitted through the leaf and misses the chlorophyll molecules, or
- more than half of the light that strikes the chlorophyll consists of wavelengths which cannot be used in photosynthesis (such as green or ultraviolet light), or
- the reactions of photosynthesis themselves are inefficient, losing much energy as heat

Primary productivity: the 0.5–1% of incident light energy that is converted into chemical energy and fixed by producers in photosynthesis is called the **gross primary productivity** (**GPP**). Some of the GPP is required by the plant for **respiration** (e.g. generating ATP for the active uptake of ions). Subtracting respiration from GPP gives the **net primary productivity** (**NPP**). The NPP is the actual rate of production by the producers (autotrophic plants) and is important because it represents the energy or biomass available:

- for the new growth of the plants, and
- to all the other trophic levels in the ecosystem

Exam tip

You should be aware that productivity of ecosystems has two components: **primary productivity** — the production of new organic matter by plants — and **secondary productivity** — the production of new organic matter by consumers. You must be able to distinguish between the two.

Exam tip

You may be asked about reasons for the inefficiency of light absorption by plants. Make sure you read the question carefully. If the question requires reasons for the inefficiency of light absorption once light has reached the vegetation, then 'absorption by the atmosphere' is not a valid answer.

Reasons for the reduction in energy at progressive trophic levels: when energy is transferred from producers to primary consumers, and to each trophic level thereafter, the efficiency is only about 5–20%. This is because:

- some of the material is not consumed, either because it is inaccessible or it is unpalatable or inedible (e.g. plant roots cannot be eaten by grazers, some plants may be spoiled by animal droppings, the hooves of a wildebeest are not eaten by a lion, the shell of a snail is not eaten by a song thrush)
- some of the material is not digested, so is not absorbed and appears in the faeces (e.g. skin and bones in the droppings of a barn owl, cellulose is not easily digested and so is egested in the faeces of herbivores)
- some of the material ends up as a waste product of metabolism and is excreted (e.g. urea in urine)
- many of the materials are used in respiration to generate ATP for active processes in the organism (e.g. ATP used in muscle contraction)

Secondary productivity: for animals, **energy budgets** can be constructed from the quantities of energy consumed (**C**), remaining as the **net secondary productivity** (**NP**), released in respiration (**R**) and leaving the animal as urine (**U**) or faeces (**F**). The equation for NP (in units of energy per unit time, e.g. $kJ\,y^{-1}$) is:

$$NP = C - (R + U + F)$$

For livestock (e.g. cattle), it is the net secondary productivity, NP, that represents the energy available for human consumption.

The transfer of energy through an ecosystem and the efficiency of this transfer between trophic levels is shown in Figure 37.

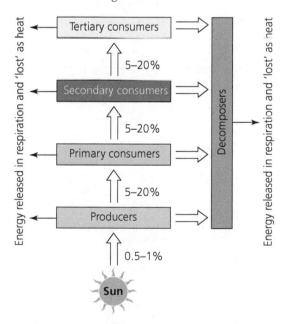

Figure 37 The transfer of energy through an ecosystem

Exam tip

You must know that the initials GPP stand for 'gross primary productivity', while NPP represents 'net primary productivity', and you must be able to distinguish between them (NPP = GPP – respiration).

Knowledge check 58

Why is the net primary productivity (NPP) often used to compare the efficiencies of different ecosystems?

Knowledge check 59

The amount of energy consumed (C) less the amount egested as faeces (F) represents the amount of energy absorbed (A). Rewrite the equation for net secondary productivity (NP) to include the amount of energy absorbed (A).

Knowledge check 60

Arctic ecosystems most often have only three trophic levels, while tropical ecosystems frequently have five levels. Explain why.

The range of 5–20% for the efficiency of energy transfer through the consumer trophic levels is a guide only and, indeed, is often summarised simply as 'the 10% rule'. The actual figure for the percentage efficiency of energy transfer will depend on the nature of the organisms, for example:

■ herbivores (primary consumers) generally have a low percentage efficiency, e.g. 13% for a grasshopper (herbivore) and 27% for a wolf spider (carnivore), because of the difficulty in digesting the cellulose in plant material, so there are relatively high losses via the egestion of faeces

■ endotherms (mammals and birds) have a low percentage efficiency (e.g. mammals have a figure of only 1–4%) because they generate heat internally to maintain a relatively high body temperature, which means their metabolism is maintained at a high rate but there are relatively high energy losses through respiration

The percentage efficiency of energy transfer between trophic levels is calculated as:

$$\text{efficiency of energy transfer} = \frac{\text{energy in one trophic level}}{\text{energy in previous trophic level}} \times 100\%$$

Productivity has been determined for an entire ecosystem — Silver Springs in Florida. The energy in gross production, net production after respiration and the energy going to the decomposers for this ecosystem are shown in Figure 38 (the unit of measurement is $\text{kJ m}^{-2}\text{y}^{-1}$).

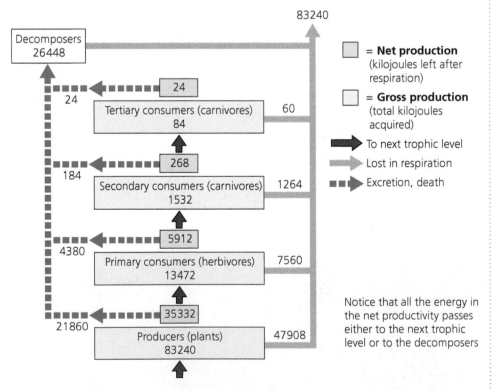

Figure 38 Gross and net productivities in the Silver Springs ecosystem

Knowledge check 61

Why do carnivores have a much higher secondary productivity than herbivores?

Exam tip

You may be asked to calculate the percentage efficiency of energy transfer between trophic levels, so you must practise using the equation opposite.

Knowledge check 62

Using the data in Figure 38, calculate the percentage efficiency of energy transfer from producers (as GPP) to primary consumers.

Look at the figures carefully. It is evident that there is less energy at each successive trophic level, but note other important findings:

■ More energy is used in respiration than is available as net production for the next trophic level — respiration generates energy for all the active processes requiring ATP in the organisms and the energy is then 'lost' as heat.

■ More of the energy from net production passes to the decomposers than to the next trophic level of the grazing food chain — for example, in a field of grass, some of the grass will have been trampled by large grazers or will have been spoiled by animal faeces and so is only available to the decomposers, while leaves may simply die.

Some of the energy in an ecosystem may be lost by the 'emigration' of material out of the ecosystem and be gained by the 'immigration' of material into it.

The implications for agriculture

The loss of energy in food webs has important implications when it comes to the production of food and our position as consumers.

Since more energy is available to primary consumers than to secondary consumers, more energy will be available to humans eating plant material than if animal products are eaten. A vegetarian diet will support many more people than a diet of meat, and in many heavily populated developing countries meat is seldom eaten. However, this does not mean that all animal husbandry is wasteful of resources. For example, in the upland areas of Northern Ireland the soil and climate will not support a crop plant which humans can eat. In these areas it makes ecological sense to raise animals, such as sheep, which can tolerate the poor-quality vegetation growing there, and then to eat these animals and/or their dairy products.

Where plant crops are grown, productivity can be increased by:

■ use of fertilisers, so that the plants have more ions, e.g. the nitrogen (N) in nitrates is used in the production of proteins
■ use of pesticides, e.g. herbicides will prevent competition from non-crop plants, while insecticides will reduce damage from insect pests

There are concerns about overuse of fertilisers and pesticides.

Where animals are farmed, productivity can be increased by:

■ feeding the animals on high-protein foods and high-energy foods such as silage (silage is made by cutting grass and enclosing it in plastic, where it is fermented by bacteria)
■ keeping animals, such as chickens and pigs, in warm conditions, so that less energy is used to generate heat
■ keeping animals in confined conditions, so that less energy is used in movement

There are ethical issues here. Some people think that it is cruel to keep animals in crowded and confined conditions where movement is restricted. Some consider that as long as the animals are kept comfortable, with sufficient food and water, then the benefit of producing food cheaply outweighs any other concern.

Knowledge check 63

Explain why a vegetarian diet can support more people.

Summary

- Light energy is trapped by producers (plants) in photosynthesis and is the external energy source for most ecosystems. Only a very small percentage of the total light energy available is utilised by producers.
- Energy is passed through food chains, which are linked with other chains as food webs.
- Energy is passed from producers to primary consumers (herbivores), to secondary consumers (carnivores), to tertiary and, perhaps, to quaternary consumers. Each level in the food chain is called a trophic level.
- Decomposers (mostly bacteria and fungi) break down the dead organic matter from producers and consumers, and play an important role in nutrient recycling.
- There are three types of ecological pyramid: pyramids of numbers, pyramids of biomass and pyramids of energy. Pyramids of energy are the most useful and show that approximately 90% of the energy transferred to the next trophic level is lost as waste matter or heat.
- Transfer of energy to carnivores is more efficient than to herbivores; endotherms (mammals and birds) have a very low efficiency of energy transfer.
- Productivity is a measure of how efficient organisms are at locking up energy in organic molecules in their body cells.
- Gross primary productivity (GPP) is all the biomass produced (or energy 'trapped') by plants (producers) per square metre per year. Since some energy is used in respiration, this leaves the plants with a net primary productivity (NPP).

    GPP = NPP + respiration

- Secondary productivity is the rate of production of biomass (or gain in energy 'trapped') by animals (consumers) in an ecosystem:

    net secondary productivity = amount consumed – amount lost in faeces – amount lost in urine – amount used in respiration

Nutrient cycling

Energy is transferred through ecosystems; it is not recycled. This is simply because in all energy conversions heat energy is released and this eventually radiates out into space. However, the elements in matter are recycled. Each element spends part of its time in complex organic molecules and part in simple, inorganic substances in the abiotic part of an ecosystem.

The carbon cycle

Plants (producers) absorb carbon dioxide (CO_2) from the atmosphere (or as hydrogen carbonate ion, HCO_3^-, from water) and synthesise carbohydrates during **photosynthesis**. From the carbohydrates produced, and using atoms in absorbed ions, the autotrophic plants produce lipids, proteins and nucleic acids, all of which contain carbon. These molecules are **consumed** by the animals (consumers) in the food chain, entering their bodies following digestion.

Exam tip

Remember energy in an ecosystem is not recycled. There must be an external energy source (usually sunlight) since energy is lost as heat as it passes along food chains. Meanwhile, there is a finite amount of matter available and so elements that make up organic organisms must be recycled when they die.

Both the plants and the animals release carbon dioxide via their **respiration**. Decomposers (most bacteria and fungi) use dead plants and animals for food. Some of this is used in respiration, releasing carbon dioxide.

Fossil fuels such as oil, coal, peat and gas contain carbon. When they are **burned**, carbon dioxide is released back into the atmosphere.

These processes are summarised in Figure 39.

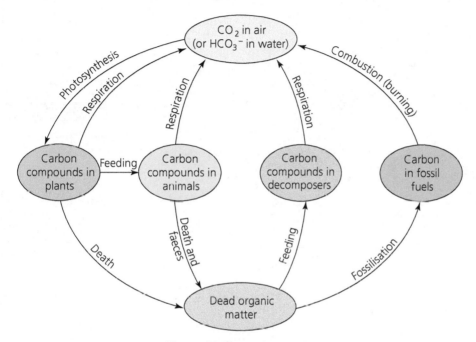

Figure 39 The carbon cycle

The nitrogen cycle

Plants (producers) absorb nitrate ions (NO_3^-) from the soil (or water, if the plants are aquatic). The plants use the nitrogen in these ions to synthesise amino acids, and so proteins, and nucleotides, and so nucleic acids. These molecules are consumed by the animals (consumers) in the food chain and enter their bodies following digestion. Undigested matter is egested as faeces.

Animals produce nitrogenous waste products, ammonia and urea, which are excreted.

Decomposers (decomposing bacteria and fungi) use dead plants and animals, faeces and urea for food. The protein (and their amino acids) and nucleic acid (and their nucleotides) are broken down and the nitrogen released as ammonia (or ammonium ion, NH_4^+).

The ammonia (NH_3) is used by **nitrifying bacteria** and converted (oxidised) to nitrate ion (NO_3^-). Nitrates are released when the bacteria die.

This description covers the basis of the nitrogen cycle (see Figure 40), except that other processes occur to either add to or reduce the nitrogen available to plants:

Knowledge check 64

By what process does carbon in the atmosphere become available to organisms in a community?

Exam tip

Many students find the nitrogen cycle difficult. You should appreciate that nitrogen is an element which is a constituent of some inorganic ions and some organic molecules: ammonium (NH_4^+) and nitrate (NO_3^-) ions are inorganic; amino acids (and so proteins) and nucleotides (and so nucleic acids) are organic. The term 'nitrogen' may be used in questions to cover the total nitrogen in an ecosystem, whether in inorganic or organic form.

Knowledge check 65

In what form do plants absorb nitrogen from their environment?

Knowledge check 66

By what processes do some bacteria convert (a) organic matter into ammonia, (b) gaseous nitrogen into ammonia and amino acids, (c) ammonia into nitrates, (d) nitrates into gaseous nitrogen?

- Nitrogen-fixing bacteria add to the nitrogen available to plants — nitrogen-fixing bacteria, some species of which (e.g. *Rhizobium*) inhabit the root nodules of legumes (e.g. clover and beans), are able to utilise gaseous nitrogen to synthesise organic nitrogen-containing compounds (i.e. amino acids).

- Denitrifying bacteria reduce the nitrogen available to plants — these bacteria, living in waterlogged and oxygen-deficient soils, use nitrate and convert it to gaseous nitrogen which is returned to the atmosphere.

Exam tip

You may be asked to reproduce or complete a diagram of the nitrogen cycle in your exam. Study Figure 40 and copy it until you can draw the cycle without reference to the figure.

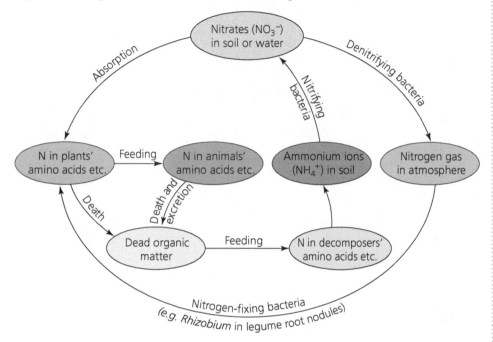

Figure 40 The nitrogen cycle

Knowledge check 67

Which of the bacteria involved in the nitrogen cycle respire anaerobically and so favour waterlogged soils?

In farm ecosystems, removing crops or grass (for silage) removes some of the primary production so that less is available to any grazing food chain. It also means that nitrogen-containing compounds within the plants are removed so that the soil becomes depleted in nitrate unless this is added in organic or inorganic fertilisers.

Summary

- Carbon is found in all organic molecules.
- In the carbon cycle:
  - CO_2 is removed from the air and fixed into organic molecules in photosynthesis
  - CO_2 is replaced in the air when any organism respires or when fossil fuels are burned
  - organic molecules are passed from plants to animals by feeding and assimilation
  - organic molecules are passed from plants and animals to decomposers, which feed on the dead remains of organisms.
- Nitrogen is found in amino acids (proteins), nucleotides (nucleic acids) and ATP.
- In the nitrogen cycle:

  - nitrate (NO_3^-) ions are absorbed from the soil by plants and used to make amino acids, proteins, etc.
  - animals eat plants, digest the plant protein and use their amino acids to make animal protein
  - when the plants and animals die, decomposers release ammonium (NH_4^+) ions from nitrogen-containing molecules
  - nitrifying bacteria oxidise the ammonium ions to nitrates
  - nitrogen-fixing bacteria (free in the soil or in root nodules of legumes) reduce nitrogen gas to ammonium ions and hence to amino acids.

Questions & Answers

The examination

The A2 Unit 1 examination contributes 24% to the final A-level outcome. The paper lasts 2 hours 15 minutes and is worth 100 marks. In Section A (82 marks) all the questions are structured, though with a variety of styles. In Section B (18 marks) there is a single question, possibly with several parts, to be answered in continuous prose.

Examiners construct papers to test different assessment objectives (AOs). In the A2 Unit 1 paper the approximate marks allocated for each assessment objective (AO) are:

AO1   Knowledge and understanding 31

AO2   Application of knowledge and understanding 43

AO3   Analysis, interpretation and evaluation of scientific information, ideas and evidence 26

As you can see from the above weightings, the application of your understanding is given even greater prominence at A2.

Skills assessed in questions

The questions in this section test the different assessment objectives. While some questions assess straightforward knowledge and understanding (AO1), most will require you to *apply* your understanding in unfamiliar situations (AO2), and there will be those that ask you to evaluate experimental and investigative work (AO3).

A2 units include some synoptic assessment, involving links between different areas of biology and including concepts from the AS units.

Since mathematical skills are an important element of biology, the questions include a variety of calculations and graphical work.

Quality of written communication, including accurate use of scientific terms, is assessed in Section B.

About this section

This section consists of questions covering the range of topics within the Content Guidance. Following each question, there are answers provided by two students of differing ability. Student A consistently performs at grade A/B standard, allowing you to see what high-grade answers look like. Student B makes a lot of mistakes — ones that examiners often encounter — and grades vary between C/D and E/U.

Each question is followed by a brief analysis of what to look out for when answering the question (shown by the icon ⓔ). All student responses are then followed by comments (preceded by the icon ⓔ). They provide the correct answers and indicate where difficulties for the student occurred, including lack of detail, lack of clarity, misconceptions, irrelevance, poor reading of questions and mistaken meanings of examination terms.

In using this section try the questions before looking at the students' responses or the comments, which you can then use to mark your work. Check where your answers might have been improved.

Read the question carefully

There are two aspects to this:

■ Pay attention to the command terms used in each question — for example, 'describe' or 'explain' — and make sure you respond appropriately.

■ The stem of a question may provide information needed to answer the question. This is particularly the case in a question presenting an unfamiliar situation. Think about how this information can help you to construct or focus on a relevant answer.

Think carefully before you begin to write. The best answers are short and relevant — if you target your answer well, you can get many marks for a small amount of writing. Don't ramble on and say the same thing several times over, or wander off into answers that have nothing to do with the question.

Use the CCEA website

The CCEA Biology specification is available from www.ccea.org.uk. Apart from the subject content, you must familiarise yourself with the mathematical skills shown in section 4.7 and the command terms used in examinations (such as explain, describe and suggest) shown in Appendix 1. The website will also allow you to access past papers and mark schemes.

Section A Structured questions

■Homeostasis and the kidney

Question 1 Filtration and reabsorption in the kidney

The diagram below shows part of a nephron, and its blood supply, where filtration and the initial stage of reabsorption take place. The figure also shows the forces involved in the filtration process. The colloidal (solute) potential in the capsule is zero, since proteins are not filtered.

Blood (in glomerulus)
Hydrostatic pressure = 8 kPa
Colloidal potential = –4.3 kPa
(due to proteins)

Filtrate (in capsule)
Hydrostatic pressure = 2.4 kPa

(a) (i) Describe the structure of the filter through which the filtrate passes. (3 marks)

(ii) Calculate the effective filtration pressure. Show your working. (2 marks)

(b) (i) The hydrostatic pressure of the blood in the capillaries surrounding the proximal tubule falls to 3.3 kPa. Assuming other values remain the same, explain why water is reabsorbed here. (2 marks)

(ii) Explain how substances other than water are reabsorbed from the proximal tubule. (3 marks)

(iii) Describe how the cells lining the proximal tubule are adapted for the reabsorption of substances. (2 marks)

(c) (i) Apart from the proximal convoluted tubule, where else is water reabsorbed? (2 marks)

(ii) As water is reabsorbed, the concentration of urea increases. The table below shows the amount of water and urea in both the filtrate and urine. Calculate how many more times urea is concentrated in urine than in the filtrate. Show your working. (3 marks)

	Mean daily amount of	
	water/dm^3	urea/g
Filtrate	180	55
Urine	1.5	35

Total: 17 marks

🅔 A number of parts in this question are assessing your mathematical skills (AO2): in (a) (ii) and again in (b) (i), where you need to understand the equation for water potential and its components (this is a synoptic link with AS Unit 1). In (c) (ii), you should be aware that there are alternative ways of tackling this and that the calculation is more a matter of common sense than of mathematical ability. The other parts are relatively straightforward, requiring an understanding of kidney function (AO1).

Questions & Answers

Student A

(a) (i) The endothelium of the glomerular capillaries is made up of cells with pores through them, ✓ while the inner lining of Bowman's capsule is made up of podocytes with elaborate foot processes between which there are filtration slits. ✓ The effective barrier, which prevents the passage of molecules above a particular molecular mass, is the basement membrane, a layer between the glomerular endothelium and the podocytes. ✓

(ii) Water potential in glomerulus = 8 – 4.3 kPa = 3.7 kPa ✓
Effective filtration pressure = 3.7 – 2.4 kPa = 1.3 kPa ✓ a

(b) (i) The water potential in the blood capillaries is 3.3 kPa less 4.3 kPa, the colloidal potential of proteins in the blood, i.e. –1 kPa. ✓ This is less than the water potential of the filtrate, which is 2.4 kPa, so water moves out of the proximal tubule along an osmotic gradient. ✓

(ii) Glucose and amino acids are reabsorbed entirely by active transport. ✓ Ions are reabsorbed partly by diffusion, as water is reabsorbed, and partly by active transport. ✓ Small proteins which may have been filtered are reabsorbed by pinocytosis. ✓

(iii) The cuboidal endothelial cells of the proximal tubule have microvilli on the lumen side and infoldings on the side nearest to the capillaries so that the surface area for reabsorption is increased. ✓ The cells possess numerous mitochondria to supply ATP for active transport. ✓ a

(c) (i) The loop of Henle ✗ and the collecting duct assuming antidiuretic hormone has been released to make it permeable. ✓

(ii) In filtrate, the concentration of urea is 55 ÷ 180 = 0.306 ✓
In urine, the concentration of urea is 35 ÷ 1.5 = 23.333 ✓
Overall factor of concentration is 23.333 ÷ 0.306 = 76.3 ✓ a

🄮 **16/17 marks awarded** a These are full and well-worded answers except for (c)(i) where Student A should have noted that water was absorbed in the descending limb of the loop.

Student B

(a) (i) The basement membrane is the actual barrier to filtration. ✓ a

(ii) The difference in the hydrostatic pressures is the effective filtration pressure. ✓ This is 8 – 2.4 = 5.6 kPa. b

(b) (i) Water is reabsorbed by osmosis along a water potential gradient. ✓ c

(ii) Glucose and other solutes are reabsorbed by active transport. ✓ Some proteins which are small enough to be filtered are taken up by pinocytosis. ✓ d

> **(iii)** The proximal tubule cells have microvilli which increase the surface
> area on which protein carriers are sited. ✓ Further, there are many
> mitochondria to supply the ATP required for active transport. ✓
>
> **(c) (i)** Water is also reabsorbed in the descending limb of the loop of Henle ✓
> and in the distal convoluted tubule and collecting duct. ✓
>
> **(ii)** 7.14 ✗

9/17 marks awarded This is correct for 1 mark but the question asked for the
layers through which the filtrate passes — this includes the glomerular endothelium
and the podocytes of the capsule. Student B has failed to take account of the
colloidal potential due to proteins within the glomerulus. Calculating the difference in
hydrostatic pressures — part of the answer — is awarded 1 mark. While the student
notes a water potential gradient, no attempt has been made to calculate the water
potential within the capillaries. Students will be expected to use their understanding
from the AS units. This is fine for 2 marks, but detail on the uptake of ions (part
diffusion, part active transport) is missing While no mention has been made of the
infoldings on the capillary side, two clear adaptations are given for 2 marks. This is
sufficient for both marks, since the question asked only for the sites of reabsorption.
This is incorrect. It is possible that the student made some correct arithmetic
operations but since no working has been shown, no marks can be awarded.

Question 2 Osmoregulation

(a) The diagrams below (not drawn to scale) show the nephrons from the
kidneys of three different mammals. Explain how nephron structure allows
these mammals to be adapted to their habitat. (3 marks)

Beaver

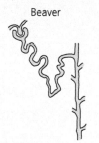

Dog

Camel

(b) Drugs that cause an increase in urine production are called diuretics. The effect of two diuretic drugs on urine production, in comparison to a control not given a drug, is shown in the graph below.

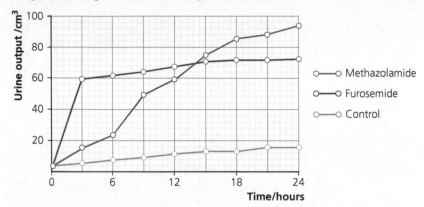

(i) Diuretic drugs may be used to control blood pressure. Explain why furosemide might reduce blood pressure and be a more effective treatment than methazolamide. **(4 marks)**

(ii) Furosemide inhibits carrier proteins involved in the active transport of sodium ions in the ascending limb of the loop of Henle. Explain why this drug increases urine output. **(3 marks)**

(c) Caffeine acts as a drug, inhibiting the release of ADH. Describe and explain the effect of caffeine on the production of urine. **(4 marks)**

Total: 14 marks

ⓔ This question is asking you to apply your understanding of osmoregulation (AO2). You must read the information and study the graph carefully before attempting to answer the questions. Also, answers will need to respond to the questions asked and not just repeat something that you have learned. To answer part (a), you should know that the beaver lives by rivers, while the camel is desert-living.

Student A

(a) The longer the loop of Henle, the more water will be reabsorbed from the collecting duct, as this creates hypertonic conditions in the medulla through which the ducts pass. ✓ The beaver has a short loop of Henle and produces a dilute urine since it can easily drink to replace any water lost. ✓ The camel needs to conserve water and so needs to produce a concentrated urine in which little water is lost. ✓ ⓐ

(b) (i) Compared to the control, furosemide increases urine output, ✓ and the subsequent loss of water from the blood reduces blood pressure. ✓ Furosemide has a more immediate effect than methazolamide, ✓ but levels off, whereas methazolamide causes continued water loss which may reduce the blood pressure too much. ✓

(ii) Stopping sodium ions from moving out of the ascending limb into the interstitial fluid causes less water to be withdrawn from the descending limb. ✓ There is a reduced concentration of salts towards the apex of the loop of Henle, so the water potential is not as low there. ✓ Less water is reabsorbed from the collecting ducts, so more is excreted. ✓ a

(c) A greater volume ✓ of dilute urine will be produced. ✓ This is because ADH makes the collecting ducts more permeable to water ✓ so less is reabsorbed and more is lost in urine. ✓ a

e **14/14 marks awarded** a Full marks for detailed answers throughout.

Student B

(a) The longer the loop of Henle, the greater the salt concentration in the medulla. ✓ This happens because sodium and chloride ions are moved out of the ascending limb, causing water to be reabsorbed from the descending limb. With a long loop of Henle and higher concentration of salts in the medulla, more water can be reabsorbed from the collecting ducts. a

(b) (i) Furosemide increases the amount of urine produced, and so fluid is lost from the body. ✓ Methazolamide continues to increase urine output, and so too much water may be lost from the body. ✓ b

(ii) If Na^+ ions are prevented from moving out of the ascending limb, then less water will be reabsorbed from the descending limb. ✓ The reduced salt concentration in the deeper regions of the medulla ✓ will mean that water reabsorption from the collecting ducts will be reduced. ✓ c

(c) Urine production will increase ✓ because ADH affects the permeability of the distal convoluted tubules and collecting ducts. d

e **7/14 marks awarded** a Everything that the student says is correct, but it does not answer the question which is asking about how the loop of Henle allows the 'mammals to be adapted to their habitat'. b Two statements are rewarded here. However, Student B has failed to compare the effect of the diuretics to the control — urine output may also have increased in the control. Also, the student fails to identify the immediate effect of furosemide compared with methazolamide. c Student B has a clear understanding of the operation of the loop of Henle, as indicated in the answer to part (a). 3 marks awarded. d This answer lacks detail. For example, there is no mention of the effect on the concentration of the urine. Even worse, it is never sufficient to say that there is an effect without indicating whether the effect is an increase or a decrease. Only 1 mark awarded.

■ Immunity

Question 3 The immune response to viral infection

In 2009, a new strain of influenza spread worldwide. The flu virus attacks the epithelial cells of the respiratory tract. Within these, new viruses are formed and are then released to infect further epithelial cells. The diagram below shows a virus-infected epithelial cell and free viruses that have been released.

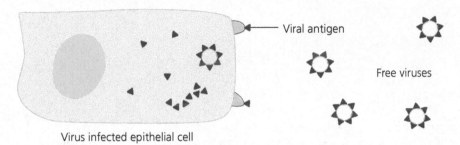

(a) Explain the role of B lymphocytes and T lymphocytes in the defence of the body against the virus infection. (6 marks)

Some people, in whom the flu symptoms were severe, were admitted to hospital. Healthcare personnel, who were required to work with these patients, received injections to provide protection:

■ Group A were given antiserum.
■ Group B were given a vaccine.

The graph below shows the antibody level in the blood plasma of the two groups over a period of 40 days after injections were given.

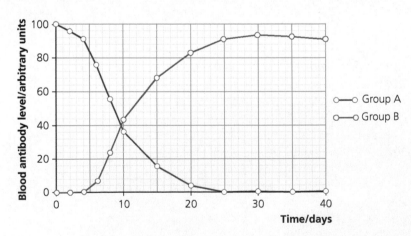

(b) Explain the shape of the curve for each of the two groups. (4 marks)

To maintain an effective level of immunity against the virus, the antibody level in the blood must be more than 36 arbitrary units.

(c) Group A healthcare staff initially attended to the infected patients. When would you expect them to be replaced by healthcare workers from Group B? Explain your reasoning.

(2 marks)

Total: 12 marks

ⓔ Part (a) is straightforward recall, though with 6 marks available you must supply a lot of detail. The other parts require you to apply your understanding. Read the introductory information and study the graph before you read the questions.

Student A

(a) *B lymphocytes*: binding of an antigen on the surface of a free virus and the complementary receptor site of a specific B lymphocyte causes an antibody-mediated response. ✓ The sensitised B-cell divides mitotically to produce a clone of cells which develops into either memory B-cells or plasma cells. ✓ Plasma cells synthesise and secrete antibodies which are specific and bind to the antigens of the viruses. ✓ The response may involve agglutination or opsonisation (✓) followed by phagocytosis.

T lymphocytes: binding of the viral antigen and the complementary receptor of a particular T lymphocyte initiates mitotic division ✓ and the resultant cloned cells develop into one of four types of cell, i.e. helper T-cells, killer T-cells, memory T-cells and suppressor T-cells. ✓ Killer T-cells produce perforins which produce perforations in the surface membrane of the infected host cell. ✓ Helper T-cells activate plasma cells and killer T cells, and stimulate the activity of phagocytes such as macrophages. (✓) ⓐ

(b) Group A workers receive antibodies in the serum, ✓ though antibody levels fall as they denature and are not replaced. ✓ Group B workers only start to produce antibodies after four days, ✓ as it takes time for B-cells to be sensitised, and for plasma cells and antibodies to be produced. ✓ ⓑ

(c) At 10 weeks. ✗ Because Group A workers are effectively immune up until this time and Group B workers become immune just after this time. ✓ ⓒ

ⓔ **11/12 marks awarded** ⓐ Six (and more) correct points are included in a good, well-worded account. ⓑ A detailed answer rewarded with 4 marks. ⓒ This is careless — the student has read the time axis as weeks instead of days. However, the reasoning is sound for 1 mark.

Student B

(a) *B lymphocytes*: the B lymphocyte is attracted by a specific viral antigen. ✗ The B-cell responds by dividing to produce a clone, the cells of which include plasma cells, and memory cells are formed. ✓ Plasma cells produce the antibodies ✓ which then destroy the free viruses.

T lymphocytes: a specific T lymphocyte has a receptor which has the same shape as the antigen. ✗ The activated T-cell divides to produce a clone of cells which differentiate to form four cell types: T-suppressor, T-killer, T-helper and T-memory. ✓ The T-killer cells engulf the infected cells. ✗ ⓐ

(b) The antiserum given to the workers in Group A contains antibodies. ✓ The workers in Group B given the vaccine do not start to produce antibodies until B lymphocytes are activated and plasma cells produced. ✓ b

(c) Group A staff have 36 arbitrary units of antibody up until 10 days, ✓ at which time Group B antibodies increase above this critical level. ✓ c

ⓔ **7/12 marks awarded** a This lacks detail and much is simply wrong: antigens are not attracted to lymphocyte receptors, they collide by chance; the receptor does not have the same shape as the antigen, the shape is complementary; killer T-cells are not phagocytes. Three valid points are made. b Again, this lacks detail, e.g. Student B could have noted that the graph shows there was a delay of four days until antibodies were produced in Group B workers. c This is correct for 2 marks.

Question 4 Transplant surgery and antibiotics

(a) Early work on transplant surgery was carried out by Medawar and Gibson on burns victims during the Second World War. The burned area was covered as quickly and completely as possible with fresh skin, to prevent infection. Experimenting with autografts from a patient's own body and allografts from donors, they found that, although both autografts and allografts initially healed successfully, allografts were rejected within weeks. When a second allograft from the same donor was attempted, the graft was rejected within a few days.

    (i) Explain why the allografts were rejected.   (3 marks)

    (ii) Explain why a second allograft from the same donor was rejected much more quickly.   (3 marks)

    (iii) State two ways in which the risk of tissue rejection might be reduced.   (2 marks)

(b) It is usual that patients undergoing surgery receive a course of antibiotics. An early antibiotic used was penicillin. This acts by loosening the links between peptidoglycan molecules in bacterial cell walls, so that, in taking up water osmotically, the bacteria burst.

    (i) Explain why penicillin is harmful to neither viruses nor human cells.   (1 mark)

    (ii) Explain why antibiotic resistance has become a major health concern.   (3 marks)

    (iii) Suggest two reasons why new antibiotics have not been discovered over the past 30 years.   (2 marks)

Total: 14 marks

ⓔ In part (a), you will probably need to read the introductory passage twice to fully comprehend the early work on skin grafts. You are not required to know the term 'allograft', you can work it out. Part (b) is more straightforward, though in (i) you will need to draw on your understanding of cellular and viral structure from AS Unit 1.

Student A

(a) (i) The allograft contains non-self antigens ✓ stimulating the sensitisation of T lymphocytes ✓ and the production of killer T-cells ✓ which cause the rejection of the allograft.

(ii) Because the second allograft from the same donor will contain the same non-self antigens ✓ and immediately be recognised by the memory T-cells produced following the application of the first graft. ✓ These divide rapidly to produce an even bigger crop of killer T-cells. ✓

(iii) The use of X-rays to stop the production of lymphocytes ✓ and the use of immunosuppressant drugs. ✓ a

(b) (i) Because only bacteria possess a cell wall. ✗

(ii) Many bacteria have evolved resistance to particular antibiotics which can no longer be used to treat bacterial infections. ✓ For example, a strain of *Staphylococcus aureus* is now methicillin-resistant (MRSA) ✓ and causes many deaths annually in patients who have undergone surgery. ✓ The use of bactericidal hand-wash is important in preventing its spread. (✓)

(iii) Pharmaceutical companies do not find it economical to invest the huge amounts necessary to discover new antibiotics. ✓ Many of the microbes which may contain antibiotics are difficult to culture. ✓ a

ⓔ **13/14 marks awarded** a Full marks for detailed answers, with the exception of (b) (i), where the answer provided is biologically incorrect — a mark would have been awarded if the student had said 'peptidoglycan cell wall'.

Student B

(a) (i) The allograft from a donor would contain foreign markers not found in the patient. ✓ These are recognised by a specific T lymphocyte ✓ which divides mitotically to produce a variety of cells, including killer T-cells ✓ which destroy the foreign tissue.

(ii) Because memory T-cells are produced ✓ as well as killer T-cells.

(iii) X-rays ✓ and certain drugs which inhibit cell division. ✓ a

(b) (i) Because antibiotics don't work against viruses. ✗ b

(ii) Antibiotic resistance develops from a few bacteria which possess a feature which protects them from the action of the antibiotic. For example, penicillin-resistant bacteria have an enzyme which breaks down penicillin. Any bacterium that survives an antibiotic treatment can then multiply and pass on its resistance. Antibiotics need to be judiciously used since overuse, or inappropriate use, increases the chance of resistant strains of bacteria developing. ✓ c

(iii) The vast majority of soil microbes, which might produce antibiotics, cannot be cultured in the laboratory. ✓ Drug companies are unwilling to make the financial outlay required to find new antibiotics. ✓ d

e **9/14 marks awarded** a Student B appears to have a good understanding of tissue rejection, but has failed, in (ii), to give sufficient detail. 6 marks awarded for part (a), but the student loses marks by not responding to an allocation of 3 marks for (ii). b This is not answering the question, simply restating information in the question. No mark awarded. c Much of this is not relevant since the question is about health concerns, not about how resistance evolves. Only the last statement is worthy of a mark. d Both statements are correct for 2 marks.

■ Coordination in plants

Question 5 Auxin and phytochromes

(a) The graph below shows the effect of applying different concentrations of auxin on the growth of lateral buds and shoots.

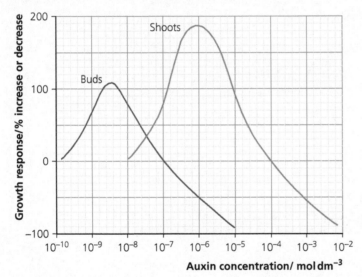

(i) Describe the effects of different auxin concentrations on the growth of shoots. (2 marks)

(ii) Suggest how a growing shoot inhibits the growth of side branches in a plant. (3 marks)

(b) (i) Complete the following passage about phytochrome.

Within leaves, the pigment phytochrome (P) exists in two interchangeable forms, P_{660} and P_{730}. P_{660} absorbs _____ coloured light, while P_{730} absorbs _____ coloured light. During the day P_____ is converted to P_____. In the dark this process is reversed. (3 marks)

(ii) The diagram below shows the flowering responses of a short-day plant and a long-day plant when exposed to a particular light regime. Explain the different flowering responses for short- and long-day plants in terms of P_{660} and P_{730}.

(3 marks)

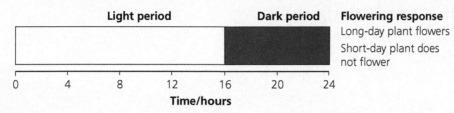

Total: 11 marks

e You are unlikely to have seen the graph in (a) before. Don't panic if this happens — this type of question is testing your AO2 skills. In (a) (i), you will need to refer to figures in the graph. In (a) (ii), 'suggest' implies that you are not expected to know the answer but should be able to work it out from the information supplied and what you already know about auxins. Part (b) (i) tests recall of the phytochrome system (AO1), but if you can remember that visible light varies in wavelength between 400 nm and 700 nm, then you can work it out. (b) (ii), about the control of flowering, requires more thought and the ability to express ideas clearly.

Student A

(a) (i) Auxin concentrations between 10^{-8} and 10^{-4} moldm^{-3} stimulate shoot growth, with a maximum increase at 10^{-6} moldm^{-3}. ✓ At high concentrations, above 10^{-4} moldm^{-3} shoot growth is inhibited. ✓ **a**

(ii) The concentration promoting shoot growth is in the region of 10^{-7} to 10^{-5}. ✓ At these concentrations bud development is inhibited. ✓ **b**

(b) (i) red, ✓ far-red, ✓ 660 to 730. ✓ **c**

(ii) The SDP does not flower because the dark period is less than the critical length and so P_{730}, which inhibits its flowering, ✓ is not sufficiently converted to P_{660}. ✓ An LDP, however, needs a short period of darkness, as here, to allow accumulation of P_{730} and so stimulate flowering. ✓ **d**

---

e **10/11 marks awarded** **a** Student A has given a full description of the trend and made specific reference to values on the graph, scoring both marks. **b** Two valid points are made, but for a third mark the student should have used previous understanding, about auxin being produced within the apical meristem and then diffusing down the shoot. **c** Three correct answers. **d** The student has produced a well-worded answer showing understanding of the relevant levels of phytochrome for this light regime and of the different responses of long-day and short-day plants to relatively high levels of P_{730}. All 3 marks scored.

Student B

(a) (i) The auxin concentration causes a rapid increase in growth up to a maximum and then growth is inhibited. ✗ a

(ii) Shoots grow in quite high concentrations of auxin, but buds are inhibited by these concentrations. ✓ b

(b) (i) red, ✓ blue, ✗ 660 to 730. ✓ c

(ii) The LDP flowers as there is enough light to convert the P_{660} to P_{730}. ✓ whereas the SDP doesn't flower as the period of time for darkness isn't long enough to convert P_{730} back to P_{660}. ✓ d

ⓔ **5/11 marks awarded** a The answer is expressed as though the x-axis showed time — for example, using the word 'rapid' and 'then'. See Student A's answer for a better way of expressing these points. b This lacks detail and fails to make any specific reference to critical values in the graph. c While phytochrome is a blue pigment, it absorbs at the red end of the spectrum. P_{730} absorbs far-red light. 2 marks scored. d The student understands the interconversions of the two forms of phytochrome in this light regime and, while not precisely stating so, clearly understands that P_{730} stimulates flowering in LDPs. However, a clear statement that the accumulated P_{730} inhibits flowering in SDPs is lacking. It is important to understand that P_{730} is the active form, whether stimulatory or inhibitory. 2 marks scored.

■Coordination in mammals

Question 6 Neurones

(a) The diagram below shows a neurone.

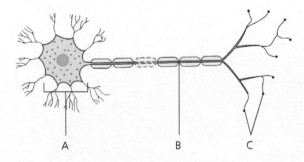

    A   B   C

  (i) Name the type of neurone shown.   (1 mark)

  (ii) Name the features labelled A, B and C.   (3 marks)

(b) The table opposite shows the speed of conduction of myelinated and non-myelinated axons in different animals.

Animal	Type of axon	Axon diameter/μm	Speed of conduction/ms^{-1}
Crab	non-myelinated	40	8
Squid	non-myelinated	700	30
Cat	myelinated	18	100
Frog	myelinated	16	35

(i) What conclusions might be made from the information provided in the table?　(3 marks)

(ii) The cat has a body temperature maintained at 37°C, while the frog's temperature rarely rises above 15°C. Suggest how the difference in body temperature might explain the difference in the speed of conduction.　(3 marks)

Total: 10 marks

ⓔ Part (a) is simple recall of neurone structure. Part (b) is more discriminating. In (i), make use of the data from the table to support any conclusions. In (ii) you will need to use your understanding from AS to work out an appropriate answer. The command term 'suggest' means that you need to think about how your understanding can be applied to this situation.

Student A

(a) (i) Motor neurone. ✓

(ii) A = centron; ✓ B = node of Ranvier; ✓ C = synaptic knobs. ✓ ⓐ

(b) (i) Myelinated axons have a faster conduction speed than non-myelinated axons. ✓ Even the slowest speed for a myelinated axon, at 35 ms^{-1} for a frog, is faster than the fastest speed for non-myelinated, at 30 for the squid giant axon. ✓ In non-myelinated axons, a large diameter allows a faster conduction speed. ✓

(ii) For an impulse to occur an action potential must be evoked. An action potential occurs as Na$^+$ ions diffuse into the axon, causing it to be depolarised. ✓ Ion diffusion will take place much more rapidly at a higher temperature, ✓ as heat increases the kinetic energy of the ions. ✓ So in the cat, ion diffusion at 37°C will be faster, the axon more quickly depolarised and an action potential passed along the axon length more rapidly. ⓑ

ⓔ **10/10 marks awarded** ⓐ All correct. ⓑ Well-phrased answers for full marks.

Student B

(a) (i) Motor neurone. ✓ ⓐ

(ii) A = cell body, ✓ B = node, ✗ C = synaptic knob. ✓ ⓑ

(b) (i) Myelinated fibres conduct more quickly than non-myelinated. ✓ ⓒ

(ii) Impulses are initiated when the axon is depolarised to cause an action potential. Depolarisation relies on ion diffusion, ✓ which is more rapid at higher temperature. ✓ ⓓ

ⓔ 6/10 marks awarded **ⓐ** Correct. **ⓑ** It is not sufficient to say just 'node' — you must know the full term, node of Ranvier. **ⓒ** This is incomplete and has made no use of the data to support a conclusion. 1 mark only. **ⓓ** There is no reference to the kinetic energy of ions, though there is sufficient here for 2 marks.

Question 7 Synapses

(a) The colour-enhanced transmission electronmicrograph below shows a synapse between a nerve ending and a striated muscle fibre (neuromuscular junction).

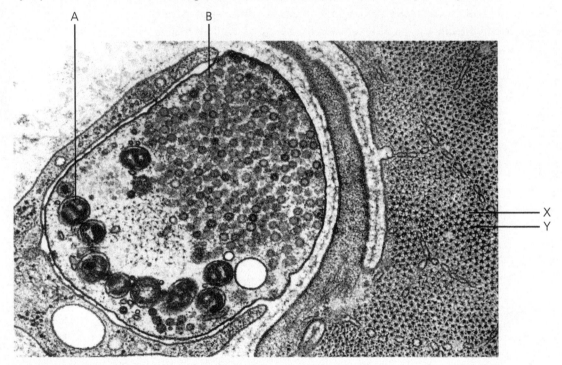

(i) Identify the organelles A and B in the synaptic knob. (2 marks)

(ii) Identify the thin filament X and the thick filament Y in the muscle fibre. (2 marks)

(iii) Name the transmitter chemical in the neuromuscular junction. (1 mark)

(b) Beta blockers are drugs that bind with receptors on the sino-atrial node of the heart, preventing noradrenaline from binding. Explain the effect of beta blockers on the heart rate. (2 marks)

(c) GABA (gamma-aminobutyric acid) is a neurotransmitter in some brain synapses. Alcohol is a drug which binds with GABA receptors and mimics the effect of GABA.

(i) Explain the effect of alcohol on the membrane potential of the post-synaptic membrane. (2 marks)

(ii) Explain the effect of alcohol on transmission in the post-synaptic neurone. (1 mark)

(iii) Suggest one consequence of alcohol on overall behaviour. (1 mark)

(d) Describe three roles of synapses. (3 marks)

Total: 14 marks

ⓔ Part (a) should be quite straightforward, even though you need to identify features in a TEM of a neuromuscular junction. In parts (b) and (c), you will need to be careful in reading the information provided; notice that in one part the drug blocks the receptors, while in the other it mimics the effect of the neurotransmitter. In part (d), you will need to provide three distinct and well-worded answers.

Student A

(a) (i)   A = mitochondrion, ✓ B = synaptic vesicle. ✓

(ii) X = actin, ✓ Y = myosin. ✓

(iii) Acetylcholine, ACh. ✓ ⓐ

(b) The sympathetic nerve secretes noradrenaline, which stimulates the SAN so that waves of excitation occur more rapidly. ✓ Beta blockers stop noradrenaline stimulating heart rate so it stays slow. ✓ ⓑ

(c) (i)   If alcohol mimics GABA, it will cause the post-synaptic potential to become more negative. ✓ The membrane hyperpolarises, creating an IPSP. ✓

(ii)   This makes it less likely for an action potential to occur and if this happens, no impulse is fired in the post-synaptic neurone. ✓

(iii) Inhibition at synapses in the brain means that alcohol will increase reaction times and make people less cautious. ✓ ⓑ

(d) To allow transmission between adjacent neurones. ✗ Ensures that transmission is in one direction only, as receptors occur only on the post-synaptic membrane. ✓ Allows control of actions through a combination of stimulation and inhibition, resulting in restraint as well as excitation. ✓ ⓒ

ⓔ **13/14 marks awarded** ⓐ All correct. ⓑ Full marks for detailed answers. ⓒ The first statement is not a role of synapses, since some are inhibitory. The following statements are valid, earning 2 marks.

Student B

(a) (i)   A = synaptic vesicle, ✗ B = mitochondrion. ✗ ⓐ

(ii) X = actin, ✓ Y = myosin. ✓ ⓑ

(iii) ACh. ✓ ⓒ

(b) They stop noradrenaline ✓ from speeding up the heart rate. ✓ ⓓ

(c) (i)   Alcohol blocks the GABA receptors. ✗ GABA normally makes the post-synaptic membrane potential more negative. ✓ ⓔ

(ii) Impulses are more readily transmitted. ✗ ⓕ

(iii) Causes erratic behaviour. ✗ ⓖ

(d) Guarantee that impulses pass unidirectionally. ✓ Protect muscles from overstimulation as neurotransmitters cannot be secreted continuously. ✓ To delay transmission between adjacent neurones. ✗ ⓗ

ⓔ 8/14 marks awarded **ⓐ** These are the wrong way round — even if the internal structure of the organelles in the TEM is unclear, you should know that mitochondria are bigger than vesicles. No marks awarded. **ⓑ** Correct for 2 marks. **ⓒ** ACh is given the mark here, but don't assume that abbreviations will always be rewarded. **ⓓ** This gets both marks, but only just — the student says that beta blockers stop the heart beating faster and links this to the action of noradrenaline. **ⓔ** Student B has misread the question — alcohol 'mimics' GABA on the receptors and so will bring about the same action. However, for knowing the effect of GABA on the membrane potential, 1 mark is awarded. **ⓕ** Student B is still thinking that alcohol blocks the GABA receptors — no mark given. **ⓖ** 'Erratic' is not precise enough to be worthy of a mark. **ⓗ** The first two statements are correct for 2 marks. The third statement is a disadvantage, not a role of synapses.

Question 8 The eye

(a) The eye is able to control the amount of light entering and accommodate (focus) that which enters onto the retina.

   (i) Describe the operation of the iris in bright light.   (2 marks)

   (ii) Describe the accommodation of the eye for a near object.   (3 marks)

(b) The graph below shows the density of rods and cones across the retina of a human eye.

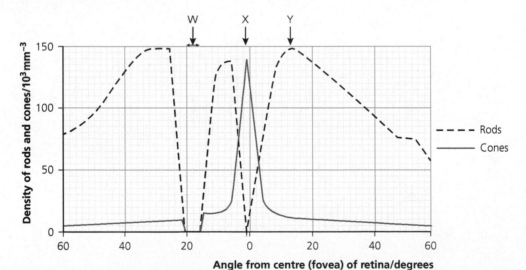

   (i) Explain why cones provide visual acuity at X but not at Y.   (2 marks)

   (ii) Explain why rods cannot provide visual acuity, even at Y.   (1 mark)

   (iii) Explain why there are no rods or cones in the region labelled W.   (1 mark)

(c) There are advantages in the retina containing a mixture of both rods and cones.

   (i) Explain how rods provide vision when the light intensity is low.   (2 marks)

   (ii) Explain how cones provide colour vision.   (2 marks)

Total: 13 marks

ⓔ This question should be easy if you have a thorough understanding of the eye. You may not have seen the graph in part (b) before, but its interpretation should be relatively straightforward. Be careful about your phrasing and that you provide complete answers to the questions asked.

Student A

(a) (i) In bright light, the iris restricts the entry of light, so preventing damage to the rods and cones. It does this by contraction of circular muscles and relaxation of radial muscles, ✓ which makes the iris bigger but constricts the pupil. ✓

(ii) When the object is near, light rays are divergent and so the lens needs to be more convergent. ✓ The circular muscles in the ciliary body contract, ✓ releasing any tension on the suspensory ligaments, so that through its elasticity the lens adopts a spheroid shape. ✓ a

(b) (i) At X, the cones are densely packed into the fovea. The cones are close together and each synapses to a single neurone, so that lots of impulses are sent to the brain — a high-resolution image is detected by the brain. ✓ At Y, there are few cones and these lie further apart. ✓

(ii) Because they display retinal convergence. They are densely packed here, but light falls on many rods synapsing with a single neurone and so points of detail cannot be separated. ✓

(iii) The neurones lie over the inner surface of the retina and unite to form the optic nerve, which leaves the eye at this spot. ✓ b

(c) (i) Many rods connect onto one bipolar neurone and so have an additive effect. ✓ c

(ii) There are three types of cone, each with a different iodopsin sensitive to blue, green and red. ✓ d

---

ⓔ **11/13 marks awarded** a Well-expressed answers. Indeed, in (i), Student A has done more than just describe how the iris works but explained the advantage of pupil constriction in bright light. b Full marks for detailed answers. c The answer is correct for 1 mark. However, a complete answer would note that a combination of rods together produces the required generator potential or that there is summation of the amount of transmitter substance released into the synapse with the neurone of the optic nerve. d This is correct but does not explain how other colours are perceived — that stimulation of combinations of cone types sends a pattern of impulses to the brain to give the perception of a particular colour. 1 mark only scored.

Student B

(a) (i) In bright light the pupil becomes smaller, ✓ so that less light enters the eye and protects the retina from damage. a

(ii) The rays of light from a near object are divergent. ✓ The ligaments relax, ✗ making the lens fatter and more refracting. ✓ b

---

(b) (i) Because there are more cones present at X than at Y. ✓ c

(ii) A bipolar neurone is stimulated by a number of rods. ✓ d

(iii) This is the blind spot. ✗ e

(c) (i) They have higher sensitivity since the pigment in rods is broken down at low light intensity. ✓ f

(ii) There are three types of cones, some sensitive to red light, some to blue light and some to green light. ✓ g

e 7/13 marks awarded a The first statement is correct for 1 mark. However, Student B has not explained how the pupil is constricted but instead talked about the advantage of constriction. b There are two correct points. Note that the ligaments are simply connective tissue transferring any tension in the wall of the eyeball through to the lens — it is the contraction of the circular muscle in the ciliary body which releases this tension. c This answer is worthy of 1 mark. A second mark requires some understanding of the synapsing of each cone to an individual neurone and the ability to discriminate points close together. d This answer is just sufficient to gain 1 mark. e Blind spot is the name given to the region, but it doesn't explain the lack of photoreceptor cells — it is the region through which the optic nerve traverses the retina. f This is an alternative answer to that of Student A, but again, it lacks the detail that a generator potential would more readily be produced. g This answer is correct, but again does not explain the perception of other colours such as purple or yellow. 1 mark only awarded.

Question 9 Muscle contraction

(a) The transmission electronmicrograph (colour enhanced) below shows part of a striated muscle fibre. Identify the features labelled A–D.

(4 marks)

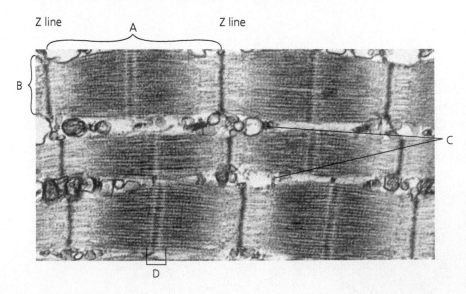

(b) The diagram below shows two cross-sectional views of the feature labelled B.

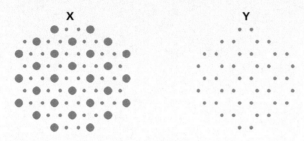

(i) Explain the views provided by X and Y. (2 marks)

(ii) Which of the views would be more prominent in a contracted muscle fibre? Explain your answer. (1 mark)

(c) Describe the sliding filament theory of muscle contraction. (4 marks)

Total: 11 marks

ⓔ Since you will not have seen this electronmicrograph before, part (a) will involve an application of your understanding of striated muscle structure (AO2). In part (b), you will need to interpret the transverse views illustrated in the diagram. Part (c) tests your knowledge of muscle contraction (AO1) — for a mark tariff of 4 marks you must provide a detailed answer.

Student A

(a) A = sarcomere, ✓ B = myofibril, ✓ C = T-tubules of the sarcoplasmic reticulum, ✓ D = H zone. ✓ ⓐ

(b) (i) X is taken through the dark end of the A band, i.e. not through the H zone where there are only thick filaments. ✓ Y is taken through the I band where there are only thin filaments. ✓

(ii) X, since when the muscle contracts the light bands and H zones become shorter, while the region of overlap increases as the thin actin filaments are drawn inwards over the thick myosin filaments. ✓ ⓐ

(c) The release of Ca⁺⁺ ions into the sarcoplasm causes binding sites on the actin filaments to become available. ✓ The myosin heads rotate, resulting in the movement of the actin filaments over the myosin rods. ✓ ATP is used to detach and re-orientate the myosin heads. ✓ These then re-attach further along the actin filament so that the process is repeated and the myofibril shortens. ✓ ⓐ

ⓔ **11/11 marks awarded** ⓐ Correct and full answers for maximum marks.

> **Student B**
>
> **(a)** A = sarcomere, ✓ B = muscle fibre, ✗ C = sarcoplasmic reticulum, ✓
> D = light band. ✗ ⓐ
>
> **(b) (i)** View X is a transverse section of the thick and thin filaments, ✗ whereas
> view Y is a transverse section of the I band. ✓ ⓑ
>
> **(ii)** View X because when the muscle contracts the thick filaments become
> more prominent than the thin filaments. ✗ ⓒ
>
> **(c)** Myosin heads attach to thin actin filaments in the presence of calcium
> ions. ✓ The orientation of the myosin heads changes, pulling the actin
> filaments over the thick myosin filaments. ✓ This process requires ATP
> expenditure. ✗ ⓓ

ⓔ **5/11 marks awarded** ⓐ B is an organelle — myofibril — within the muscle
fibre. The answer of C is allowed, though Student A's answer is more complete.
D is not the light band — the Z line runs through the light (I) band; it is the H
zone within the dark (A) band. 2 marks awarded. ⓑ The first point does not
explain why both thick and thin filaments are apparent — it is a section through
the A (dark) band, though not the central H-zone component of it. The second
is correct for 1 mark. ⓒ There is no mark for guessing X and, in any case, the
reasoning is incorrect. The region of overlapping filaments increases as the
filaments slide over each other when the muscle contracts. ⓓ The first two
statements are correct. However, Student B then says that 'this' process,
suggesting rotation of the myosin head, requires ATP usage — it is the release
and re-orientation of the myosin head which needs ATP. Since there is no further
detail, only 2 marks are scored in what was an easy topic if only the student had
worked harder to learn it.

■ Ecosystems

Question 10 Population dynamics

(a) Great tits (*Parus major*) are small birds which pair and form territories
within which they nest, producing a clutch of about ten eggs. A
population of great tits in Marley Wood near Oxford was intensely studied
over 70 years.

The graphs opposite show results relating to the numbers of eggs laid per
nest and subsequent hatching failure.

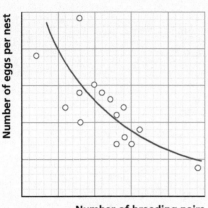
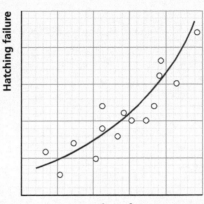

(i) Describe and suggest an explanation for the relationship shown in each graph. (4 marks)

(ii) After an especially severe winter, mortality in the great tit population was high and numbers plummeted. Explain how the population would recover during the subsequent summer. (2 marks)

(b) In a population of robins (*Erithacus rubecula*), ringing experiments, over many years, allowed the mortality of different age classes to be studied. The results of the study are shown in the table below.

Age/years	Number surviving at beginning of age interval	Number dying in age interval
0–1	140	98
1–2	42	20
2–3	X	16
3–4	6	Y
4–5	3	0
5–6	2	0
6–7	2	0
7–8	2	0
8–9	1	1

(i) Determine the values for X and Y in the table. (2 marks)

(ii) A pair of robins has an average clutch size of five eggs and produce two clutches per year. Using the information in the table, calculate the number of young birds produced by a pair of robins which might be expected to survive their first year. Show your working. (2 marks)

(iii) Robins are reproductively mature at one year old. Such early maturation is a feature of r-selected species. State one other characteristic of r-selected species. (1 mark)

Total: 11 marks

ⓔ In part (a), to describe each relationship, simply put into words the trend shown, remembering that the variable on the y-axis depends on the variable on the x-axis. When asked to 'suggest an explanation', remember that you are not expected to know the answer but should be able to apply your understanding to put forward a valid reason. In part (b), you will need to study the table of data carefully and gain an understanding of the survival of robins as they get older. Remember that in mathematical problems, as in (b) (ii), you are asked to show your working because 1 mark may be available for a correct arithmetic operation even if the final answer is incorrect. Part (b) (iii) is simple recall.

Student A

(a) (i) When there are more breeding pairs, there are fewer eggs laid per nest; ✓ territories will be smaller, so less food is available for egg production. ✓

When there are more eggs laid per nest, few eggs hatch; ✓ eggs in more dense clutches may have less yolk to sustain the developing embryo. ✓

(ii) After the severe winter, there would be fewer breeding pairs but with larger territories, ✓ so that more eggs would be laid per breeding pair. ✓ The population would recover as egg production increases. **a**

(b) (i) X = 22, ✓ Y = 3. ✓

(ii) Mortality in the first year is 98 ÷ 140 = 70%. ✓

70% of 10 young robins = 7, meaning that while 7 die in their first year, 3 survive. ✓

(iii) r-selected species produce large numbers of young. ✓ **a**

ⓔ 11/11 marks awarded **a** All correct for full marks.

Student B

(a) (i) The bigger the population of great tits, the fewer eggs are produced in each nest. ✓ This is because the adults cannot get enough food for the young. ✗ When there is greater hatching failure, there are more eggs per nest. ✗ The adults may not be able to incubate all the eggs properly when there are lots in the nest. ✓ **a**

(ii) If the population was to plummet then there would be fewer pairs surviving to breed. ✓ Fewer breeding pairs will have sufficient food to produce large clutches ✓ and so the population recovers as more young are produced. **b**

(b) (i) X = 42 − 20 = 22, ✓ Y = 6 − 2 = 4. ✗ **c**

(ii) Survival rate in the first year is 42 ÷ 140 = 0.3. ✓

Two clutches of five = 10 young, which means that 10 × 0.3 survive = 3 birds survive. ✓ **d**

(iii) They are easily adaptable. ✗ **e**

ⓔ 7/11 marks awarded **ⓐ** Regarding the first graph, the relationship is correctly described but the reasoning is not valid since eggs have not yet hatched! For the second graph, the description is poorly worded and Student B has got the dependency of the variables the wrong way round. However, the reasoning is valid and an alternative to that given by Student A. **ⓑ** Both points are valid for 2 marks. **ⓒ** X is correct, Y is not. This looks like a slip and shows the value of always checking your calculations. **ⓓ** This is correct and an alternative solution to that of Student A. 2 marks awarded. **ⓔ** This is too vague. There is a long list of r-selected features. No mark awarded.

Question 11 Competition and species interaction

Duckweed is a small plant that grows on or near the surface of ponds. It reproduces asexually, producing leaf-like structures called fronds.

(a) In an investigation, two species of duckweed, *Lemna gibba* and *Lemna minor*, both floating species, were grown separately and also together in tanks of pond water. The growth of each species was measured by counting the number of fronds. The results of the investigation are shown in the graphs below.

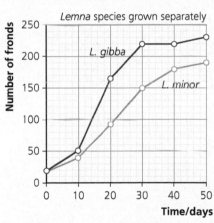
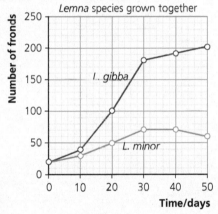

(i) State three abiotic factors that should be controlled in order to ensure valid results in this investigation. (3 marks)

(ii) Determine the weekly growth rate for each species when grown separately from 0–4 weeks. (2 marks)

(iii) What is the evidence that the two species are in competition when grown together? (1 mark)

(iv) Suggest an explanation for the outcome of competition between the two species. (2 marks)

(b) In a second investigation *Lemna minor* and a third species, *Lemna trisulca*, were grown separately and also together. The results of the investigation are shown in the following graph.

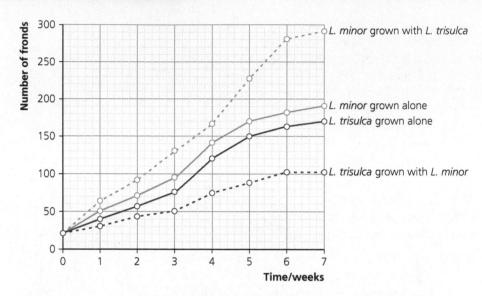

While *L. minor* grows on the surface of pond water, *L. trisulca* grows under the surface. Compare the growth of each species when grown separately and when grown together, and comment on the nature of the interaction taking place. (4 marks)

Total: 12 marks

(e) Study the graphs carefully to gain an understanding of the interactions illustrated. You may have met inter-specific competition graphs before and you can use this understanding to answer part (a). Part (a) (i) is relatively straightforward, testing your understanding of an appropriate control of variables in an investigation of plant growth. In (a) (ii), you need to read values carefully from the graphs, paying attention to the units of time. The use of the term 'suggest' in (a) (iv) means that while you are not expected to know the answer, you can work it out. Part (b) presents a more novel situation and so is more difficult — the command term 'comment on' invites you to describe, draw conclusions and offer explanations about the nature of the interaction.

Student A

(a) (i) Light intensity, ✓ temperature, ✓ nutrients. ✓

(ii) For *Lemna gibba*: $(210 – 20) ÷ 4 = 47.5$ week^{-1} ✓

For *Lemna minor*: $(140 – 20) ÷ 4 = 30$ fronds week^{-1} ✓

(iii) When growing together the number of fronds produced by each species is considerably less than when each is grown separately. ✓

(iv) The two species may have been competing for light. ✓ It may be that *L. gibba* is able to float above *L. minor* and block light from reaching it. ✓ [a]

(b) When grown separately they grow at a similar rate. However, when grown together *L. minor* grows more rapidly, ✓ while *L. trisulca* does not grow as quickly. ✓ When interacting, *L. minor* benefits from the presence of *L. trisulca*. ✓ [b]

e **11/12 marks awarded** **a** Correct and full answers for 8 marks. **b** There are three points worthy of marks. However, there are other points that may have been made, e.g. that this is not competition, it is a +/– interaction, and *L. minor* is obtaining some substance from *L. trisulca* beneficial to its growth.

Student B

(a) (i) Minerals, ✓ pH, ✓ absence of other plants. ✗ **a**

    (ii) For *Lemna gibba*: $210 \div 4 = 50.5$ week^{-1}

       For *Lemna minor*: $140 \div 4 = 35$ fronds week^{-1} } ✗ ✓ **b**

    (iii) They are not reaching so high a level. ✗ **c**

    (iv) *L. gibba* grows at a faster rate than *L. minor* and so outcompetes it. ✓ This may be because *L. gibba* takes up mineral ions and uses them more efficiently. ✓ **d**

(b) In normal competition, the stronger competitor still loses out when grown in mixed culture. However, *L. minor* appears to benefit from the interaction. ✓ This is not mutualism as *L. trisulca* still loses out. ✓ This must therefore be an exception to inter-specific competition. **e**

---

e 7/12 marks awarded **a** Competition from other plants would be a biotic factor. 2 marks scored. **b** 1 mark is awarded for reading the figures at 28 days (4 weeks) from the graph. However, the student has forgotten to subtract the initial density of 20 in each case, so failing to score a further mark. **c** This is just too vague. What does 'they' refer to? No mark scored. **d** This is an entirely reasonable explanation and an alternative suggestion to that made by Student A, gaining both marks. **e** Two good initial points, but the student seems to have given up trying to provide further explanation.

Question 12 Predator–prey interactions and biological control

Mites are minute animals belonging to the phylum Arthropoda, class Arachnida, sub-class Acari. They are a very diverse group. While many species feed on living or dead plant material, some are carnivorous.

(a) State the taxonomic group (taxon) which lies immediately:
  - below sub-class
  - above species. (2 marks)

(b) The orange feeding six-spotted mite, *Eotetranychus sexmaculatus*, is preyed upon by the predatory mite, *Typhlodromus occidentalis*. The following graph shows the results of a laboratory study of the predator–prey relationship.

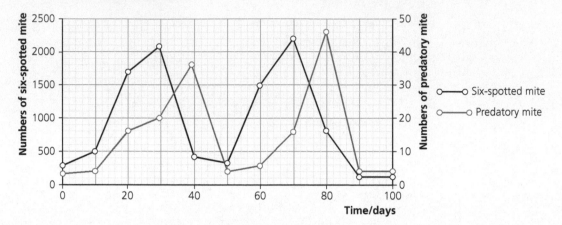

Describe and explain the results shown in the graph. (4 marks)

(c) The two-spotted mite, *Tetranychus telarius*, is a serious pest of greenhouse plants, such as tomatoes. To avoid economic damage, two methods — chemical control using dicofol and biological control employing the predatory mite *Phytoseiulus persimilis* — were tested. The graphs below show the results of an investigation comparing the effectiveness of these two methods. The arrows show when the chemical pesticide was applied and when the predator was introduced.

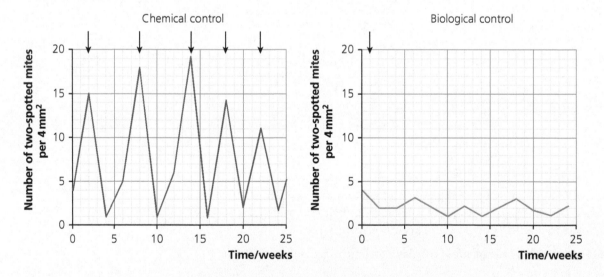

Comment on the effectiveness of the two control methods. (4 marks)

Total: 10 marks

ⓔ Part (a) tests your knowledge of taxonomy and links with AS Unit 2. While you may not have seen the graphs in parts (b) and (c) before, their analysis is not difficult, though you need to provide well-expressed and complete answers. Note the use of the term 'comment on' in (c) — expect to describe differences, offer explanations and provide an evaluation of the investigation.

Student A

(a) Family, ✓ genus. ✓ a

(b) When the numbers of the predatory mite are low, the six-spotted mite is free of environmental resistance and population growth is rapid. ✓ With more prey available for the predatory mites, they reproduce to increase in number. ✓ There is a lag of approximately 10 weeks between the peak in prey numbers and the peak in predators. ✓ With large numbers of predators, many prey are eaten and the prey numbers plummet. ✓ The decrease in prey numbers means less food for the predatory mites and their death rate increases while reproduction is reduced. (✓) The maximum number of prey mites is approximately 50 times higher than the maximum number of predators. (✓) b

(c) With chemical control, numbers of the pest mite fluctuate over the half-yearly period. ✓ This means that the chemical dicofol has to be applied each of the five times that pest numbers peak. ✓ This is because not all of the pest mites are sprayed with the chemical and some survive to reproduce. ✓ With biological control, the numbers are reduced to a level which never rises above three per $4\,mm^2$, ✓ so that, over the half year, the pest never causes economic damage. ✓ The predatory mite never eats all its prey and, if it did, it would itself be eliminated, and the pest mite might be introduced from another area. (✓) c

🅮 **10/10 marks awarded** a Both taxons are correct. b This is a complete answer, including accurate reference to the data. The maximum 4 marks gained. c A very complete analysis of the graphs with more valid statements than there are marks available. 4 marks awarded.

Student B

(a) Family, ✓ family. ✗ a

(b) The number of prey mites increases and then decreases. This pattern is repeated twice over the 100-day period. The increase in prey mites is followed by an increase in predator mites. ✓ b

(c) Chemical control is only effective at reducing the numbers of the two-spotted mite over a short period. The two-spotted mite is never completely eliminated and, being a pest, has a high rate of reproduction. ✓ The frequent application of dicofol may be because the pest mite has become immune to the chemical. ✗ Since chemicals have to be frequently applied, chemical control is probably more expensive than biological control. ✓ Biological control keeps the numbers of the pest at such a low level that it cannot cause economic damage over the period shown. ✓ c

🅮 **5/10 marks awarded** a Family is correct as the first taxon, but not for the second — family is above 'species' in the taxonomic hierarchy but not immediately above. b Student B does not explain the changes, though gains 1 mark for describing the pattern. c There are three valid points here and a fourth would have been available if Student B had used the term 'resistant' rather than 'immune', which has a totally different meaning. 3 marks scored.

Question 13 Communities and succession

(a) Distinguish between primary and secondary succession. (2 marks)

(b) With the end of the Ice Age in Ireland about 10,000 years ago, as the climate warmed, the bare rock exposed was open for colonisation by plants and animals. Several types of ecological succession took place. Over most of lowland Ireland, the climax community was deciduous forest dominated by ash and oak. Describe the features of a succession towards a climax community such as deciduous forest. (4 marks)

(c) With the retreat of the ice, thousands of lakes were left in central Ireland. These lakes underwent succession and where there was poor drainage and the ground was waterlogged, this culminated in the formation of peat bogs dominated by the moss, *Sphagnum*. The groundwater becomes acidic, bog plants are not fully decomposed and the habitat's only source of nutrients is within rainwater.

   (i) Suggest why trees are unable to survive in waterlogged soils. (2 marks)

   (ii) In central Ireland in 2003, two Iron Age bodies were found in peat bogs. Despite having been dead for over 2000 years, the bodies were well preserved with muscle, skin and hair. Suggest why the bodies had not decomposed. (2 marks)

(d) Heather (*Calluna vulgaris*) shrubs are the dominant species in some moorlands. Human activity can maintain heather moorland at an early stage of succession by periodically burning the heather through controlled firing. After burning an area, a further succession takes place, as shown in the table below.

Time after burning/ years	Appearance of heather plant	Mean % cover of heather	Relative abundance of other plant species
3		10	Many
10		90	Few
18		70	Some
25		25	Many

(i) What name is used to describe a succession which is prevented, by human activity, from reaching its climax? (1 mark)

(ii) Explain why the number of other plant species decreases between 3 and 10 years after burning. (2 marks)

(iii) As the heather plants age, they become more woody so that after 10 years there is a decline in productivity: the amount of new growth is greatest at 10 years, while at 25 years there is a high biomass but little new growth. Suggest two distinct reasons for the decrease in heather productivity after 10 years. (2 marks)

Total: 15 marks

ⓔ This question examines your understanding of community succession. Part (c) has synoptic links to AS Unit 2. In part (d), you need to study the information in the table before attempting the questions.

Student A

(a) Primary succession begins with bare rock exposed by geological activity, such as volcanoes or glaciers. ✓ Secondary succession begins with soil from which a previous community has been removed, e.g. by fire. ✓ a

(b) The bare rock is colonised by pioneer species such as moss. ✓ As a soil develops, ✓ other plant species such as grasses and shrubs become established. ✓ As the soil further develops and nutrient levels increase, conditions are changed ✓ to allow trees to grow. a

(c) (i) If the soil is waterlogged then there is no oxygen ✓ and so the roots cannot respire. ✓

(ii) Decomposers, such as bacteria, are not active, ✓ since the soil has a very low pH. b

(d) (i) Sub-climax or plagioclimax. ✓

(ii) As the young heather plants grow, they outcompete the other plant species ✓ for resources such as light or water. ✓

(iii) The number of other plant species is increasing and competing with the heather ✓ for soil nutrients. Also, as the plants age they become woody and the increased biomass means that there is a high rate of respiration ✓ so that there is little food to produce new leaves. c

ⓔ **14/15 marks awarded** a All answers correct. b Correct, except for (c) (ii) where, for the second mark, the student should have noted that a low pH (or acidic conditions) stops enzymes working. c Full marks. In part (d) (i), the term plagioclimax is correct and so is awarded the mark even though it is not given in the specification. The answer to (d) (iii) is particularly impressive.

Student B

(a) Primary succession takes place where the environment has not previously been colonised. ✓ Secondary succession takes place after the primary succession. ✗ a

(b) Small plants such as mosses act as pioneers. ✓ As these die and rock erodes, a soil is formed and humus develops. ✓ These changes allow many more plant species to colonise the area ✓ and so biodiversity increases. ✓ Eventually trees colonise the area. ⓑ

(c) (i) Waterlogged soils are anaerobic. ✓ ⓒ

   (ii) Because the soil is waterlogged. ✗ ⓓ

(d) (i) Biotic climax. ✓ ⓔ

   (ii) There is an increase in inter-specific competition, ✓ possibly for soil nutrients ✓ such as nitrates. ⓕ

   (iii) The heather needs to be burned again to increase productivity. ✗ ⓖ

ⓔ **9/15 marks awarded** ⓐ The first is correct for 1 mark; the second statement is too vague. ⓑ This is a reasonable account for 4 marks. ⓒ This gains 1 mark, but there is no mention of oxygen being required for the roots' respiration. ⓓ This is not sufficient and indeed is given in the question stem. Student B needs to explain why bacteria are not active in anaerobic and highly acidic conditions. ⓔ Correct for 1 mark. ⓕ Both points correct. 2 marks awarded. ⓖ This is not an answer to the question asked. No mark awarded.

Question 14 Ecological energetics and practical ecology

A meadow contained a diverse range of plants (e.g. grasses and meadow buttercup) and animals (e.g. grasshoppers and rabbits).

(a) Approximately $4 \times 10^6 \, kJ \, m^{-2} y^{-1}$ of light energy reaches the meadow. Of this only $2 \times 10^4 \, kJ \, m^{-2} y^{-1}$ is converted into chemical energy by the plants.

   (i) Explain why all the light energy that reaches the meadow is not converted into chemical energy. (3 marks)

   (ii) Explain why all the energy that is converted into organic compounds by the plants is not available to the herbivores such as rabbits. (2 marks)

(b) The diagram below shows the energy budget for a rabbit. (Figures are $kJ \, day^{-1}$.)

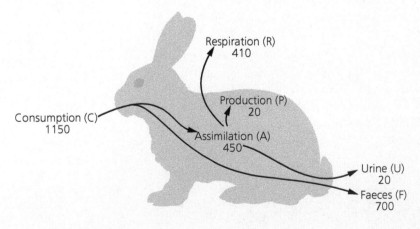

Respiration (R)
410

Production (P)
20

Consumption (C)
1150

Assimilation (A)
450

Urine (U)
20

Faeces (F)
700

(i) Write an equation for production (P) in the rabbit. Use the symbols used in the diagram opposite. (1 mark)

(ii) Calculate the percentage of energy in the food consumed that is lost to decomposers. Show your working. (2 marks)

(iii) The fox is a carnivore. In a fox, the percentage of energy lost to decomposers is 15%. Explain why the rabbit and fox figures differ. (1 mark)

(c) Describe how the population density of meadow buttercups could be estimated using a quadrat. (4 marks)

(d) Describe how the size of the grasshopper population in the meadow could be estimated using the capture–recapture technique. (4 marks)

Total: 17 marks

ⓔ Part (a) is a standard question on the inefficiency of energy transfer, while in (b) you must study carefully the energy budget for a rabbit before attempting the questions. Parts (c) and (d) test your understanding of practical work (AO3). For (c) you will need to remember ecological techniques from AS Unit 2.

Student A

(a) (i) Some of the light will pass through the plants by transmission, ✓ some will be reflected, ✓ some of the light will be of the wrong wavelength, ✓ while there is a loss of energy within the photosynthesis reactions. (✓)

(ii) Some of the organic compounds are used in respiration, ✓ some form inedible material such as roots, ✓ while others are within dead leaves which enter the decomposer food chain. (✓) a

(b) (i) P = C − F − U − R ✓

(ii) (700 + 20) ✓ ÷ 1150 × 100% ✓ = 62.6%

(iii) There is relatively more faeces produced by the rabbit as it is a herbivore and its diet has a lot of cellulose which is not fully digested. ✓ b

(c) A quadrat of size ½m × ½m is randomly positioned within the meadow. ✓ The number of meadow buttercups in the quadrat is counted ✓ and the process repeated approximately 30 times. ✓ The mean number of buttercups per quadrat is calculated. c

(d) Some grasshoppers should be caught and marked with a permanent marker on the underside. ✓ This will mean that it isn't removed and does not make the grasshoppers more obvious to a predator. They are then released back into the meadow. After they have mixed within the meadow population, say one day, further grasshoppers are caught ✓ and the number that are recaptures, that is, marked, are counted. ✓ d *e* **15/17 marks awarded** [a] More points have been provided for the maximum 5 marks. [b] All answers correct for 4 marks. [c] There are three valid points for 3 marks. A fourth point would be to describe how quadrats might be randomly positioned, or how to calculate the total number of buttercups within the meadow — multiplying the average per $0.25\,m^2$ by the area of the meadow in m^2. [d] Again, three valid points for 3 marks. However, there is no mention of how grasshoppers might be captured, e.g. use of a sweep net, and no suggestion of how the population size would be estimated, e.g. by multiplying the first and second sample sizes and dividing by the number of recaptures.

Student B

(a) (i) Some of the light is the wrong wavelength so is not absorbed by the chloroplasts. ✓ For example, green light is not absorbed — mostly only red and blue. [a]

(ii) Some of the organic compounds are used by the plants in respiration. ✓ [b]

(b) (i) $P = C - (R + U + F + A)$ ✗ [c]

(ii) 700 ✗ $\div$ 1150 × 100% ✓ = 60.9% [d]

(iii) The faeces of the rabbit contain cellulose. ✓ [e]

(c) A quadrat is randomly arranged at least 10 times in the meadow. ✓ A random arrangement of the quadrats is obtained by using random number tables to produce the coordinates for each position. ✓ The number of meadow buttercups in each quadrat is counted ✓ and an average is calculated. [f]

(d) A number of grasshoppers are captured using a sweep net, ✓ marked and released. A day later when the grasshoppers have had time to mingle with the remaining population, another sample is captured ✓ and the number of marked ones counted. ✓ The number initially marked is multiplied by the number of recaptures to get the number in the population. ✗ [g]

e **10/17 marks awarded** [a] The answer is correct for 1 mark, but with a mark tariff of 3, two other points are still required. [b] Again, only one point is provided when two were required. 1 mark only scored. [c] This is incorrect. A should have been omitted from the student's equation — it is a major part of consumption (C), i.e. that part of consumption not egested as faeces (F) and includes production (P), respiration (R) and excretion (U). An alternative equation for production includes assimilation and is $P = A - R - U$. [d] The amount of material given as going to decomposers is incorrect since it ignores the urine. 1 mark is awarded for the correct denominator and the procedure for calculating a percentage. Note that marks are awarded for correct procedures, not just for the correct answer. [e] There is just sufficient here for 1 mark, though the answer might have noted the difficulty of digesting cellulose. [f] Student B has three distinct points for 3 marks, but has failed to record the size of quadrat used or how to calculate the density of meadow buttercups within the whole meadow. [g] Again, the student scores 3 marks for three valid points, but makes no reference to a suitable marking method, while the equation for determining population size is wrong.

Question 15 The nitrogen cycle

(a) The diagram below shows part of the nitrogen cycle.

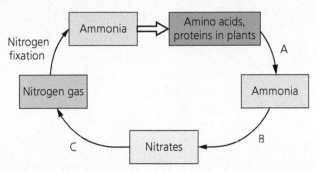

(i) Name the processes A, B and C. (3 marks)

(ii) Explain why process C is favoured in waterlogged soils. (1 mark)

(b) The pea plant is a leguminous plant. It can produce root nodules which are colonised by nitrogen-fixing bacteria as part of a mutualistic interaction between plant and bacterium.

The mass of root nodules possessed by plants in soil to which nitrogen fertiliser had been applied and in soil lacking any addition of nitrogen fertiliser was investigated. The results of the investigation are shown in the table below.

	Mass of root nodules in each of 10 plants/g
Soil with N-fertiliser applied	0.9, 0.8, 1.6, 1.0, 1.2, 0.4, 1.9, 1.1, 2.1, 1.3
Soil lacking addition of N-fertiliser	2.7, 3.2, 2.4, 3.8, 2.8, 4.1, 3.6, 2.6, 4.8, 2.3

(i) Describe and suggest an explanation for the results of the investigation. (3 marks)

(ii) Suggest a statistical analysis of the data. (3 marks)

(c) The bacteria colonising the root nodules of legumes produce nitrogenase enzymes which are responsible for nitrogen fixation. The efficiency of these enzymes is reduced in the presence of oxygen. Root nodules may appear pink because they produce the protein leghaemoglobin in response to being colonised by these bacteria. Leghaemoglobin is chemically and structurally similar to haemoglobin. Explain how the relationship between the bacteria and the legume benefits both organisms. (3 marks)

Total: 13 marks

e Part (a) should be relatively straightforward recall. Parts (b) and (c) are more difficult: you must analyse the data in (b) and read carefully the information in (c) before answering the questions.

Questions & Answers

Student A

(a) (i) A = decomposition, ✓ B = nitrification, ✓ C = denitrification. ✓

(ii) Denitrifying bacteria obtain oxygen from nitrate when oxygen gas is not available. ✓ a

(b) (i) Soil which has no N-fertiliser added has larger root nodules. ✓ Larger root nodules mean that there is more room for nitrogen-fixing bacteria and so more ammonia produced within the pea plants to compensate for the lack of nitrogen in the soil. ✓ This suggests that the pea plants respond to the low nitrogen levels in the soil by growing larger root nodules to accommodate the bacteria. ✓

(ii) The means ✓ and standard deviation of the means are calculated for each condition — soil with and without N-fertiliser added. ✓ This allows a t-test to be undertaken to determine if the difference between the means is statistically significant. ✓ a

(c) The leghaemoglobin has a high affinity for oxygen and so will associate with it. ✓ This is beneficial to the bacteria as it prevents the oxygen inhibiting the nitrogenase enzymes, ✓ allowing the conversion of nitrogen gas into ammonia. b

e **12/13 marks awarded** a All correct for 10 marks. The answers in (b) are full and particularly well worded. b 2 marks awarded — this is an excellent explanation of the role of leghaemoglobin and how this will benefit the bacterium. Unfortunately, Student A fails to state the benefit of the relationship to the plant — it obtains a source of ammonia from which it can synthesise amino acids.

Student B

(a) (i) A = decay, ✓ B = ammonification, ✗ C = denitrification. ✓ a

(ii) The bacteria which denitrify function in anaerobic conditions. ✓ b

(b) (i) When nitrogen fertiliser is added the root nodules are smaller ✓ since they do not need lots of nitrogen-fixing bacteria. ✓ c

(ii) The means are compared ✓ using a t-test. ✓ d

(c) The bacteria will produce ammonia from nitrogen gas and give some to the plant. ✓ The plant provides the bacteria with somewhere to live and will give them shelter and food, for example glucose. e

e **8/13 marks awarded** a 2 marks scored. Ammonification is another name for decomposition or decay. b Correct for 1 mark. c This is sufficient for 2 marks. For a complete answer see that of Student A. d Again this just earns 2 marks. For a third mark the student needed to refer to calculating the standard deviation for each sample or the standard deviation (error) of each mean. e 1 mark is awarded, for correctly stating that the plant would obtain ammonia from the bacteria. However, the rest is too vague (maybe an additional mark would have been awarded if the student had stated that the plant provides glucose for the bacteria's respiration), while there is no use of the information provided in the question stem.

Section B Essay questions

Question 16 Acquired immunity and immunoassay

(a) Give an account of the different ways in which immunity can be acquired. (Details of antibody- and cell-mediated responses are not required.)   (12 marks)

(b) Outline the technique of enzyme-linked immunosorbent assay (ELISA).   (6 marks)

In this question, you will be assessed on your written communication skills.

Total: 18 marks

ⓔ You must write in continuous prose since you will be assessed on your written communication skills. In both parts you should take some time to devise a plan. Notice that in part (a), you must avoid giving undue detail and ensure that your account is well balanced. The essays are marked according to the number of worthy (called indicative) points within the content and placed in one of three bands; the final mark is then based on the quality of written communication. The quality of written communication is assessed on the clarity and coherence of your writing and on the use of appropriate specialist vocabulary.

Student A

(a) Naturally acquired immunity occurs without the intervention of the medical service. The difference between the 'active' and the 'passive' forms depends on whether the person makes the antibodies themselves (actively) or gets them from another host (passively). ✓ Naturally acquired active immunity occurs when the person is exposed to a live pathogen, develops the disease ✓ and becomes immune as a result of a primary immune response. ✓ This response involves the activity of antibody-producing plasma cells (and killer T-cells), though memory cells are also produced so that immunity is long term. ✓ Naturally acquired passive immunity occurs when a mother gives her own antibodies to her baby, transferring them from her blood to the fetal blood across the placenta, or giving them to the baby in her breastmilk. ✓ The mother's antibodies provide protection to the baby against infections that the mother has encountered during her lifetime. ✓ Since the baby didn't produce the antibodies itself, protection is short term ✓ though vital since the baby's immune system takes several years to develop. ✓

In artificially acquired immunity the person is intentionally exposed to foreign antigens (actively), or given antibodies produced within another host (passively). ✓ Artificially acquired active immunity is produced by the injection of a vaccine containing the attenuated pathogen or its antigen. ✓ The antigens in the vaccine stimulate the immune system to produce antibodies and memory cells. After vaccination, if the living pathogen with the same antigens gets into the body, antibodies are either already produced or memory cells generate a rapid immune response ✓ and the pathogen is quickly destroyed without the person suffering the disease. ✓ Artificially acquired passive immunity is obtained by giving a person an injection of serum containing antibodies made by another host. ✓ Since there is no involvement of the person's own immune system, protection lasts only a few weeks. ✓ However, the neutralising effect of the antibodies is important in immediately dealing with venomous toxins, ✓ or in the case of a virulent disease, such as Ebola, when serum can be obtained from a person recovering from infection. ✓ a

(b) ELISA is used to detect a protein ✓ that is acting as a biomarker of a medical condition. ✓ The technique is based on the production of a specific antibody which will attach to the target protein, with an enzyme bound to the constant region of the antibody. ✓ The antibody is produced from a single clone of plasma cells, cultured from an isolated mouse B-cell in a laboratory — these are called monoclonal antibodies. ✓ The antibody attaches to the protein if it is present in a test sample taken from a patient. ✓ The enzyme acts on a colourless reagent, following its addition to the test sample, forming a coloured product. ✓ The intensity of the colour indicates the amount of protein in the test sample. ELISA is used to identify the proteins associated with certain cancers, e.g. PSA protein in prostate cancer, and cytokines associated with inflammation and indicative of atherosclerosis. ✓ In a similar technique, antibodies can be produced to target specific protein antigens on cancer cells and carrying, instead of an enzyme, a highly toxic anti-cancer drug, so that the drug gets delivered directly to cancer cells while doing minimal damage to other cells — an antibody with a drug attached is called a 'magic bullet'. a

Quality of written communication. b

ⓔ **18/18 marks awarded** a Succinct and precise essays. The very high number of worthy points puts the student in the top band for indicative content in both parts. b The excellence of the quality of written communication means that Student A is awarded the maximum number of marks.

Student B

(a) The immune system is vital for the defence of the body against pathogens such as bacteria, viruses and fungi. The barrier to infection, such as the skin, is the first line of defence, the action of phagocytes, such as macrophages, forms the second line, with the specific immune response representing the third line of defence.

Specific immunity is acquired when the lymphocytes respond to invading pathogens. The pathogen has non-self antigens on its surface ✓ which are recognised and initiate the activity of a specific B lymphocyte and T lymphocyte. ✓ Both types divide mitotically to produce a clone. B-cells produce antibodies that destroy the pathogen. T-cells produce T-killer cells that produce perforins that destroy the pathogen or infected host cells. Apart from these, memory cells are produced which ensure that long-term immunity is achieved. ✓ This type of immunity is natural. ✓

Immunity can also be achieved artificially. ✓ This occurs through vaccination with weakened infectious pathogens or their antigens. ✓ Again, the non-self antigen initiates the activity of specific B- and T-cells, and since memory cells are produced, immunity is long term.

> In conclusion, the immune system's defence mechanisms are well designed to allow sufficient protection for the body. If the first line of defence is broken, then the body is still protected by the second and third lines of defence. If all of these lines of defence fail, then the body suffers a serious infection or disease.
>
> Quality of written communication a
>
> **(b)** Enzyme-linked immunosorbent assay is used to detect cytokines, a small protein produced by inflamed tissue. ✓ It is also used in pregnancy tests. The pregnancy test checks for the presence of human chorionic gonadotropin (hCG) produced by the placenta. The absorbent tip of a test strip is dipped in a sample of urine. As the urine diffuses along the test strip, any hCG present binds with its specific antibody on the strip. The antibody–hCG complex diffuses further along and binds with a second antibody (complementary to the first). ✓ The second antibody incorporates an enzyme ✓ which causes a reagent to change colour. ✓ The change in colour indicates the presence of hCG and thus pregnancy.
>
> Quality of written communication b

e 10/18 marks awarded a Too much of this is irrelevant: the writing of an introduction and conclusion is unnecessary, while the question stem emphasises that 'details of antibody- and cell-mediated responses are not required'. The result is that insufficient time has been given to devising a plan with the emphasis on writing a balanced essay. The number of points for indicative content put the essay in the mid-band and with a lack of coherence though with reasonable use of technical language, a total of 6 marks are awarded for part (a). b The student has explained ELISA through the example of the pregnancy test but has not picked out sufficiently all the features of the technique. With the points given within the indicative content, Student B is placed in the middle band and, with good coherence and use of terminology, a final mark of 4 is awarded.

Question 17 Neurotransmission

(a) Describe the excitation at synapses and transmission along neurones. (12 marks)

(b) Discuss how each of the following drugs interferes with neurotransmission. (6 marks)

Drug or poison (and origin)	Effect
ω–Agatoxin (funnel web spider)	Blocks calcium receptors
Batrachotoxin (poison dart frogs)	Depolarises membrane by opening sodium channels
Morphine (opium poppy)	Blocks receptor sites on post-synaptic membrane
Organophosphates (man-made nerve gases)	Inhibit the action of cholinesterase enzyme
Tetrodotoxin (puffer fish)	Blocks sodium channels, so preventing action potentials

In this question you will be assessed on your written communication skills.

Total: 18 marks

Questions & Answers

e Part (a) requires you to outline transmission, both across synapses and along neurones. You must ensure that you cover both and that answers are detailed if you are to access all the marks available. For such a long answer, you will obviously benefit from making out an essay plan beforehand. While the question part is AO1, it is not just simple recall. You need to think carefully about the sequence of actions that will take place from the moment that an impulse arrives at a synaptic knob. Part (b) is entirely AO2 and requires you to interpret the actions of five drugs or poisons. You will have to read the information carefully in order to work this out. Quality of written communication is assessed in this question and will be based on how well you sequence your ideas and are able to draw out the links between cause and effect.

Student A

(a) Transmission occurs across synapses following the arrival of an action potential at a synaptic bulb. This causes the pre-synaptic membrane to become permeable to calcium ions, which enter the synaptic bulb. ✓ This causes the synaptic vesicles to fuse with the synaptic membrane, ✓ secreting acetylcholine into the synaptic cleft. ✓ The acetylcholine diffuses across the cleft and fuses with receptor sites on the post-synaptic membrane. ✓ This causes an influx of sodium ions, creating an excitatory post-synaptic potential (EPSP). ✓ If this reaches a threshold level, an action potential is evoked. ✓

Neurones or axons have a resting potential when they are negative on the inside. ✓ When stimulated the potential difference is reversed and an action potential occurs. ✓ The depolarised section then undergoes repolarisation and has a refractory period ✓ when no further action potentials can occur. However, a local circuit with the neighbouring section ✓ causes depolarisation and an action potential there. ✓ The result is that action potentials are propagated down the length of the axon. ✓

Quality of written communication a

(b) ω-Agatoxin 'blocks calcium receptors' and so no Ca^{2+} ions can enter the synaptic knob, meaning that the synaptic transmitter will fail to be released. ✓

Batrachotoxin 'depolarises the membrane by opening sodium channels', which means that the impulse is unable to pass along the neurone. ✓

Morphine 'blocks receptor sites on the post-synaptic membrane' so that acetylcholine cannot bind with them and set up an excitatory post-synaptic potential. ✓

Organophosphates 'inhibit cholinesterase', meaning that the ion channels are kept open and impulses continue to be fired in the post-synaptic neurone. ✓

Tetrodotoxin 'blocks sodium channels' so that action potentials are prevented.

Quality of written communication b

ⓔ 16/18 marks awarded **ⓐ** The detail and balance between synaptic and impulse transmission is good and the many valid points place the indicative content in the top band. The student expresses ideas clearly and fluently through well-sequenced sentences, while the use of biological vocabulary is good throughout, so that a total of 12 marks is awarded for part (a). **ⓑ** The actions of the first four drugs are well explained. However, the answer regarding tetrodotoxin is simply a restatement of the information in the question. The student needs to add that no impulse can be fired in the post-synaptic membrane and note that ultimately there will be paralysis or loss of sensation. Student A is placed in the mid-band for indicative content and, with consistently well-expressed answers, awarded 4 marks for part (b).

Student B

(a) Following stimulation of receptors, waves of excitation are passed along ✗ sensory neurones until they reach synaptic knobs. In the synaptic knob, Ca^{2+} ions initiate the process whereby vesicles, containing acetylcholine, fuse with the pre-synaptic membrane ✓ and pass acetylcholine via exocytosis. ✓ Acetylcholine, the transmitter substance, binds with the receptors on the post-synaptic membrane ✓ and opens up protein channels. Na^+ ions then pass through the post-synaptic membrane, causing it to be depolarised. If sufficient depolarisation ✓ occurs, an action potential is achieved and the signal is passed onwards. Once on the motor neurone, the action potential moves along the axon ✗ to the effector, where the response takes place. Both the sensory and motor neurones are myelinated, meaning that the signal is passed on with greater velocity. This is due to the signal 'jumping' from one node of Ranvier to the next. This is called saltatory transmission. ✓

Quality of written communication **ⓐ**

(b) ω-Agatoxin 'blocks calcium receptors' and so prevents synaptic transmission from being initiated, ✓ as it is Ca^{2+} that binds to receptors on the pre-synaptic membrane that start the response.

Batrachotoxin 'depolarises the membrane by opening sodium channels', which means that the post-synaptic membrane is constantly depolarised and the absolute refractory period is constant.

Morphine 'blocks receptor sites on the post-synaptic membrane' so that acetylcholine cannot attach and no depolarisation of the post-synaptic membrane takes place. ✓

Organophosphates 'inhibit cholinesterase', which means that acetylcholine is not broken down ✓ and cannot be passed back to the pre-synaptic side.

Tetrodotoxin 'blocks sodium channels' so Na^+ cannot pass into the post-synaptic membrane and so no action potentials can be achieved.

Quality of written communication **ⓑ**

e 9/18 marks awarded **a** Five appropriate points are provided. The essay lacks detail, which is virtually absent when describing the transmission of an impulse along the neurone. For example, with respect to synaptic transmission, where do the calcium ions come from? Regarding impulse transmission, there is nothing about the potential difference across the axon membrane, depolarisation of the neighbouring part of the axon membrane, recovery of resting potential or the impulse as a self-perpetuated action potential. This response is placed in the mid-band for indicative content, and quality of written communication is assessed. The student expresses ideas clearly, but the phrasing and use of biological vocabulary are somewhat limited so that a total of 6 marks is awarded for part (a). **b** The student too often repeats the information in the stem without making deductions. Still, there are three valid points and, with the assessment of quality of written communication, 3 marks are awarded.

Knowledge check answers

1 To increase the length of the tubule and so increase the surface area for reabsorption.

2 The pressure is high because the renal arteries are short and close to the heart and because the afferent blood vessel carrying blood to the glomerulus is wider than the efferent vessel.

3 **(a)** The endothelial cells are thin and contain pores, which allow through all constituents of the blood plasma. **(b)** The basement membrane is permeable to small molecules (smaller than those with a relative molecular mass of 68000). **(c)** There are numerous slits between the foot-like processes of the podocytes, which allow free flow of substances into the cavity of the capsule.

4 It is the same (because the basement membrane is permeable to glucose).

5 Some of the molecules in the nephron are waste (e.g. urea) and must be left in the fluid to be excreted. Other molecules are useful to the body and must be reabsorbed.

6 **(a)** Osmoreceptors of the hypothalamus; **(b)** pituitary gland (posterior lobe).

7 More ADH.

8 A short loop of Henle results in less water being reabsorbed but because water is readily available to beavers (they live beside or in water) they do not need to conserve it so much.

9 A parasite is an organism that lives in or on its host for all or part of its life; it causes harm and gains nutrition from its host. A pathogen is any organism that causes disease.

10 The bacteria in the intestines are adapted to the conditions there and competitively exclude pathogenic bacteria.

11 Our own cells carry antigens, just like any other cell. The antigens associated with our own cells are recognised as 'self'. This is due to a process during development that destroys any cells from the immune system that have receptors complementary in shape to the antigens on our own cells. Therefore, once the immune system is mature, it contains no cells with receptors that attach to our own antigens and stimulate a response.

12 Only one or a few B lymphocytes are specific (carry the appropriate receptor) to an antigen. Each sensitised B lymphocyte then produces many memory cells, which can respond to the antigen if met again.

13 Helper T-cells help coordinate the immune response; they activate B-cells and stimulate phagocytes. Killer T-cells attack and destroy infected host cells by injecting toxins into cells that show signs of infection.

14 If the helper T-cells are destroyed by viral action, the person will have a weak immune system and will be more susceptible to infections.

15 Bottle-fed babies do not receive the antibodies contained in the mother's milk.

16 *Differences*: B-cells respond to 'free' bacteria and viruses, while T-cells respond to infected cells, cells changed by cancer or cells transplanted from another individual; B lymphocytes divide to produce plasma cells and memory cells, while T lymphocytes divide to produce killer T-cells, helper T-cells, suppressor T-cells and memory T-cells. *Similarities*: both have specific surface proteins that match the antigens (on the surface of the bacterium or the infected cell); both exhibit a primary response (via the production of antibodies or the action of killer T-cells) and a secondary response (via the production of memory cells).

17 Memory lymphocytes for the disease would be present in the immune person and not in the non-immune person.

18 The second dose is a 'booster' to increase the number of memory cells.

19 This is active immunity because the person is making their own immune response to an antigen, producing their own antibodies and memory cells.

20 After a vaccine is administered, it takes time for a primary response to occur and for specific antibodies and memory cells to be produced. If a person is already infected, a primary response will be happening in their body anyway. Vaccines are effective only if a primary response takes place before the person is infected, so that a rapid secondary response is stimulated when the person encounters the antigen.

21 **(a)** The antibody has a binding site which is complementary only to PSA. **(b)** To remove unbound antibodies — if they remained in the well they would give a false positive result. **(c)** To react with the indicator solution (substrate) — the colour change provides a visual test indicating the quantity of antigen present.

22 Because viruses have neither cell walls nor ribosomes.

23 The bacteria would need to have appropriate mutations in each of two genes.

24 Each growth substance has a specific shape that will bind only to a specific receptor with a complementary shape on the surface membrane of particular cells.

25 Far-red light would cause the rapid conversion of P_{730} to P_{660}. The level of P_{730} would be reduced and flowering would not occur.

26 The CNS (grey matter) — motor neurones transmit impulses from the CNS to effectors.

27 Schwann cell.

28 The resting potential is determined by an unequal distribution of charged ions inside and outside a neurone, making the inside negative to the outside. This is mainly due to an excess of sodium ions (Na^+) outside.

29 During the refractory period the membrane, immediately behind that exhibiting an action potential, is unexcitable and so the impulse can only be propagated forward. The refractory period is largely responsible for the unidirectional propagation of action potentials along neurones.

30 **(a)** The impulse would be slower as it would have to make more jumps. **(b)** No impulse because the nodes are too far apart to form a circuit; too far for the action potential to jump.

31 They produce ATP for the synthesis of neurotransmitter substance (e.g. acetylcholine from choline and ethanoic acid diffusing back into the bulb).

32 Calcium.

33 **(a)** Curare will prevent further transmission (along neurones and muscle fibres) and lead to paralysis (so that breathing stops). **(b)** Sarin will lead to the continued stimulation of neurones and muscle fibres, resulting in muscle convulsions and leading to muscle fatigue (so that breathing stops).

34 It refracts light as it enters the eye so that light is focused onto the retina. The cornea refracts light more than the lens, though the lens adjusts the focusing for near and far objects.

35 The lens fattens (thickens).

36 Cones capable of colour perception operate in bright light; in dim conditions the cones cannot operate and the rods, which can operate in dim light, lack colour vision.

37 Blue, green and red.

38 Binocular vision is simply vision with two eyes. Stereoscopic vision involves the ability to combine the images formed by the two eyes to produce a single, three-dimensional image.

39 Cardiac, smooth and skeletal muscle.

40 I bands, H zones and sarcomeres become shorter, while A bands stay the same.

41 **(a)** Calcium ions unblock the binding sites on the actin filaments by moving the blocking proteins, so that myosin heads can bind onto actin filaments, initiating the formation of acto-myosin bridges and muscle contraction. **(b)** ATP is used to break the acto-myosin bridges so that the cycle of attachment, rotation and detachment can continue.

42 **(a)** Almost equal, but the reproduction rate is slightly higher. **(b)** Reproduction rate is higher. **(c)** Equal.

43 Biotic factors: competition and predation. Abiotic factors: light and mineral ions.

44 During the exponential phase, intra-specific competition is minimal because resources are in plentiful supply. However, during the stationary phase, intra-specific competition is intense because resources are limited.

45 An r-selected species.

46 Predators kill their prey to obtain nourishment, parasites do not.

47 Mutualism.

48 **(a)** It takes time for the predator to respond to changes in food supply, for example time to reproduce in response to an increased food supply. **(b)** The predator does not eat all the available prey/ energy is lost between trophic levels.

49 Competition would eventually lead to the exclusion of one species.

50 Highly species specific, long term, no environmental contamination.

51 Pioneer species, through their activities, may create a soil suitable for the next stage of succession.

52 Light.

53 Primary succession involves the introduction of plants and animals into an area that has not previously supported a community. Secondary succession refers to the reintroduction of organisms into an area previously occupied but perhaps destroyed by fire. Secondary succession usually occurs faster because it takes place where soil, often containing seeds, is present.

54 Hawks occupy both the secondary and tertiary levels.

55 Because the trees are very large and represent a huge biomass.

56 **(a)** While water is valuable it does not have an *energy value* for the consumer. **(b)** Heating above 105°C may burn the plant material, with organic material being broken down by combustion.

57 Because the producers (first trophic level — phytoplankton) are reproducing at a sufficiently high rate to support the larger biomass of consumers (second trophic level — zooplankton), i.e. the producers have a greater productivity.

58 Because NPP represents the total energy available to consumers and the whole ecosystem.

59 NP = A − (R + U).

60 Because the NPP of the tropical forest plants is much greater than the NPP of the arctic plants and so there is much more energy available to support consumers.

61 A protein-rich diet is more readily and efficiently digested; faeces contain less undigested matter as there is no cellulose in the diet.

62 $\frac{13472}{83240} \times 100\% = 16.2\%$.

63 Eating plants directly eliminates further links in the food chain and so the energy 'loss' (approximately 90% between one trophic level and the next) is reduced.

64 Photosynthesis/carbon dioxide fixation.

65 Nitrate ions (from the soil).

66 **(a)** Decomposition/ammonification; **(b)** nitrogen fixation; **(c)** nitrification; **(d)** denitrification.

67 Denitrifying bacteria.

Index

Index